CAMBRIDGE SOCIAL BIOLOGY TOPICS

Series editors
S. Tyrell Smith and Alan Cornwell

HUMAN NUTRITION

Ann F. Walker

Lecturer (Human nutrition)
Department of Food Science and Technology
University of Reading

CAMBRIDGE
UNIVERSITY PRESS

Acknowledgements

The publishers would like to thank the following for permission to reproduce photographs, diagrams and tables:
Fig 8.1 from Garrow J.S. 1983, Desirable Weight, *British Nutrition Foundation Nutrition Bulletin Number 39*, **8**, 149-155; table in appendix from *Report on Health and Social Subjects Number 15*, HMSO, reproduced with the permission of the Controller of Her Majesty's Stationary Office; quotation in preface from Pike M. 1975, *Success in Nutrition*, John Murray (Publishers) Ltd; table 1.1 from Wilson E.D. et al 1975, *Principles of Nutrition*, John Wiley and Sons Ltd, New York; photographs for plate 1 supplied by Dr P.R. Murgatroyd of the Dunn Clinical Nutrition Centre, Cambridge.

The publishers would like to thank the following for permission to adapt diagrams and tables for use in this book:
Fig 5.1 from Best C.H. and Taylor N.B. 1958, *The Living Body. A Text in Human Physiology (4th edition)*, Chapman & Hall, London; tables 6.1 and 6.5 and figure 6.1 from Davidson S. et al 1979, *Human Nutrition and Dietetics (7th edition)*, Churchill Livingstone, Edinburgh.

Cover photograph © Caroline Penn.

Published by the Press Syndicate of the University of Cambridge
The Pitt Building, Trumpington Street, Cambridge CB2 1RP
40 West 20th Street, New York, NY 10011–4211, USA
10 Stamford Road, Oakleigh, Melbourne 3166, Australia

First published 1990
Reprinted 1992
Reprinted with updated statistics 1993

Printed in Great Britain at the University Press, Cambridge

British Library cataloguing in publication data
Walker, Ann Frances
    Human nutrition.
    1. Man. Nutrition – For schools
    I. Title
    613.2

ISBN 0 521 31139 X

SE

Foreword

Message from
The Viscount Boyd of Merton
Chairman of Save the Children

'You are what you eat' is a saying with sad significance for all too many children. An inadequate diet during those vital early years makes children more vulnerable to disease and in severe cases can retard both physical and mental development. And the problems of malnutrition are not confined to the Third World.

In our work both at home and overseas Save the Children is dedicated to giving children the best possible start in life. Programmes in over 30 countries include primary health care, nutrition centres, work with disabled children and those whose only home is the street. We are helping communities to meet their childrens' basic needs with village vegetable gardens, clean water supplies and income generation schemes.

Projects in the UK are centred in inner cities, and include self-help family centres, schemes to help young offenders get off to a fresh start, support for homeless families and unaccompanied refugee children.

On behalf of everyone at Save the Children I thank Ann Walker for donating the royalties of this book to assist our work.

Preface

This book is primarily intended for use in schools, but it should find application at college level as well. Although the approach has been modified to accommodate the expected readership, the material has, to a large extent, been drawn from the various courses that I teach in the Department of Food Science and Technology at the University of Reading.

For this reason, I am grateful to nearly a generation of undergraduate and postgraduate students, both from home and overseas, for their keen interest in human nutrition and feedback on the subject matter. While human nutrition is an integral part of the subject of food science, so is food science an essential part of nutrition and for this reason some aspects of food science have found their way into this book (particularly Chapter 2 on basic chemistry of major food substances and Chapter 9 on food from production to consumption).

Another group of students whose course has influenced this monograph are those taking the course leading to the MSc in Tropical Agricultural Development. Their requirements in terms of food and nutrition education necessitate an orientation towards Third World countries, where the consequences of poor diet continue to plague humankind. Nutrition problems peculiar to the Third World have been highlighted as necessary, but especially in Chapters 8 and 9.

Considering that good nutrition for all is an aim that is still to be achieved in many parts of the World, it gives me great pleasure to donate all the royalties of the sale of this book to Save the Children. This donation is doubly inspired. Firstly by the generous nature of my late mother, Marguerite 'Daisy' Walker, who died during the preparation of this book, and to whose memory the donation will be a lasting memorial. Secondly by the resolute work of HRH The Princess Royal on behalf of the Save the Children Fund.

I have greatly appreciated the encouragement, help and support of my husband, Dr Alan Lakin, during the preparation of this book. I am also grateful to my colleague Dr Wilkie Harrigan for his expert assistance with microbiological aspects of Chapter 9.

Contents

| | Foreword | iii |
| | Preface | v |

1	**Introduction**	3
1.1	Nutrient requirements	3
1.2	Recommendations for nutrient intake	6
1.3	Dietary surveys and composition of foods' tables	8
1.4	Dietary guidelines	9

2	**Basic chemistry of major food substances**	11
2.1	Carbohydrates	11
2.2	Proteins	15
2.3	Fats	19

3	**Digestion: alimentary canal, digestion and absorption**	22
3.1	The necessity for digestion	22
3.2	The structure of the alimentary canal	22
3.3	Structure of the mucosa of the small intestine	24
3.4	Control of gut secretions	24
3.5	Digestion: carbohydrates	25
3.6	Digestion: lipids	26
3.7	Digestion: proteins	27
3.8	Absorption	28
3.9	Dietary factors inhibiting iron and calcium absorption	30
3.10	Factors increasing iron and calcium absorption	30

4	**Metabolism of carbohydrate, protein and fat**	31
4.1	Introduction	31
4.2	Carbohydrate metabolism	31
4.3	Protein metabolism	34
4.4	Fat metabolism	36

5	**Major organs of metabolism: normal and abnormal functions**	38
5.1	Endocrine glands	38
5.2	Thyroid	38
5.3	The pituitary gland and the hypothalamus	39
5.4	Liver	40
5.5	Pancreas	41
5.6	Blood glucose regulation	41

6	**Nutrient balance**	45
6.1	Introduction	45
6.2	Energy	45
6.3	Energy expenditure and energy intake	46
6.4	Protein	49

CONTENTS

6.5	Water	50
6.6	Minerals	50
6.7	Vitamins	52
6.8	Essential fatty acids	54
6.9	Dietary fibre	55
6.10	Dietary fibre: definitions, analysis and sources	57
6.11	Dietary fibre: its nutritional contribution	58
6.12	Physiological effects of dietary fibre	59

7	**Dietary requirements of various human groups**	61
7.1	Factors influencing energy and protein requirements	61
7.2	The adequacy of diet	63
7.3	Pregnancy and lactation	64
7.4	Infants and children	64
7.5	The elderly	66
7.6	Vegetarians	67
7.7	The sick, institutionalised and those with specific disorders	67
7.8	Diabetics	68

8	**Malnutrition: over- and undernutrition**	69
8.1	Overnutrition and obesity	69
8.2	Coronary heart disease	71
8.3	Effect of diet on plasma lipids	72
8.4	Undernutrition	74
8.5	Protein energy malnutrition	74
8.6	Goitre	76
8.7	Xerophthalmia	76
8.8	Anaemia	77

9	**Food from production to consumption**	79
9.1	Introduction	79
9.2	Health hazards from food	81
9.3	Food processing	86
9.4	Function of preservatives and additives	86
9.5	Effects of processing on microbiological safety of food	87
9.6	Effects of processing on the nutritional value of food	88
9.7	Conclusion	90

Further reading		92
Appendix		94
Index		96

1 Introduction

Food is eaten when we are hungry and because we enjoy it. Although foods eaten vary very widely among different individuals, depending on their economic and cultural background, the overall contribution to the diet must fulfil minimum criteria in terms of nutrients present and their quantity for maintaining health.

Nutrition in general terms is the study of the efficiency of food to nourish the body. Although we all have our own concept of what human nutrition is, it is, in fact, it is difficult to define in precise terms. The point is well made by Dr Magnus Pyke in his book *Success in Nutrition*:

'Success in nutrition can only be recognised in health, and the concept of health is a complicated one. Not for nothing is it often bracketed with happiness. Nutrition is in some respects a branch of chemical science or, more exactly, of biochemistry, that is, the chemistry of life; but it is more than this. Children cannot grow properly unless they are given the right food to eat: this is part of the science of nutrition. But there is also now good scientific evidence to show that children and young animals grow better if they are given attention and love as well as vitamins and proteins. An expert committee of the World Health Organisation has defined health, towards which good nutrition is intended to contribute, as 'complete physical, mental and social well-being and not the mere absence of ill-health and infirmity'.'

From this definition, an important difference between human and animal nutrition emerges. Human nutrition has the aim of **health** and **well-being**, while animal nutrition (farm animal at least) has more tangible aims, such as rate of growth or egg production, things which can be readily measured. Neither health nor well-being can be readily measured; indeed, no single measurement would take in the complexity implied in these terms. So, human nutritionists start with this disadvantage, and this is compounded by the ethics of experimenting with human beings. These problems mean that we have much less precise information about human nutrient requirements than we have for animals. A further complication is the enormous variability of human beings, which makes generalisations difficult.

1.1 Nutrient requirements

So how do we know about what nutrients are required by humans and what do we know about the amounts required per day? This information comes from three main types of study: body composition, nutrient deficiency and nutrient balance.

3

Whole body composition

As you might imagine this could be determined in the manner of some notorious murderers in history attempting to dispose of the evidence by acid dissolution, followed by analysis. You may not think this to be a serious proposal for a scientific investigation, but it has been done. Not by many people, it takes a brave and determined soul to undertake such work, but such people have done stirling work in providing information which may never be repeated. A summary of the results of these analyses can be seen in Table 1.1.

Table 1.1 Chemical composition of an adult human body

Element	%
Oxygen	65
Carbon	18
Hydrogen	10
Nitrogen	3.0
Calcium	1.5–2.2
Phosphorus	0.8–1.2
Potassium	0.35
Sulphur	0.25
Sodium	0.15
Chlorine	0.15
Magnesium	0.05
Iron	0.004
Manganese	0.000 3
Copper	0.000 15
Iodine	0.000 04
Cobalt	trace
Zinc	trace
Fluorine, selenium	?
Molybdenum, chromium	?

While, of course, analysis of body composition in this way is extremely crude and gives us no idea of the organic compounds present in the body, it does show the major and minor elements present. The nutrients required by the body are likely to reflect its composition, so this is a start. However, it will not give us any indication of the amounts of these nutrients required, except in broad terms.

The composition of the human body can also be expressed in terms of the protein, fat, carbohydrate, water and mineral content. Perhaps it is not surprising to learn that, if the human body is analysed at different stages of life, the composition differs (Figure 1.1). The foetus contains very little fat, compared with a baby at term. There is also a considerable change in the water content at all ages and in the mineral content of the body between the baby stage and the adult.

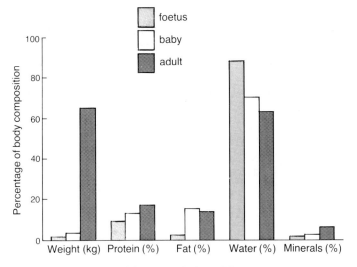

Figure 1.1 Composition of the human body at different ages

If nutrient intake is, at least to some extent, a reflection of body composition, these differences give us the first indication that the requirements of nutrients may be different at different ages of life. However, actual levels required cannot be derived from this information.

Nutrient deficiency studies

In this type of study a diet is fed lacking a particular nutrient and its effect noted. This technique was pioneered using experimental animals and it was in this way that our understanding of the importance of vitamins in nutrition was identified. Later, deficient diets were fed to humans to note the effect. From this type of study it has emerged that humans have requirements for over 40 nutrients (listed in Table 1.2), and these differ in a number of respects from animal requirements.

Table 1.2 Nutrients essential in the diet

Proteins	*Essential amino acids* – leucine, isoleucine, lysine, methionine, phenylalanine, threonine, tryptophan, valine, and histidine (infants)
	Nitrogen to synthesise non-essential amino acids
Lipids	Linoleic acid
Minerals	Calcium, phosphorus, iron, iodine, sodium, potassium, sulphur, chlorine, magnesium, zinc, manganese, copper, cobalt, fluorine (possibly molybdenum, selenium, chromium)
Vitamins	*Fat-soluble* – A, D, E, K
	Water-soluble – thiamin, riboflavin, niacin, B_6, folic acid, B_{12}, pantothenic acid, biotin and C
+ Energy	

A natural extension of deficiency studies is that, after reaching the point when deficiency symptoms are first noticed, the nutrient is added back into the diet and the amount necessary to just relieve the symptoms is noted. This gives some indication of the basic **physiological requirement** of a nutrient for that individual.

One of the most famous studies of this nature was the Sheffield study on vitamin A during World War II. Conscientious objectors were required to 'volunteer' for useful service as they were not prepared to fight. In some cases this involved taking part in nutritional studies. In Sheffield, 'volunteers' were placed on a vitamin A-free diet. It was several months before the first symptoms showed and these were recorded as the effect on dark adaptation (see Chapter 8). Vitamin A was then added back to the diet at various levels until it just relieved the symptoms.

Nutrient balance studies

In these studies the intake of a particular nutrient can be measured in relation to the output from the body. The output is usually in faeces and urine, with smaller amounts lost from other body tissues (see Section 6.4). For most purposes it is sufficient to measure loss in faeces and urine and use published values for other losses. In healthy, well-nourished adults, the intake of a nutrient should be equivalent to the output and this is the point of **nutrient balance** or **equilibrium**. If the intake is lower than required, however, the amount of a nutrient lost will be greater than the amount taken in, as for all nutrients there is a daily loss (this has been called the **obligatory loss** in the case of protein). Under these conditions the person is said to be in **negative nutrient balance**.

Subjects in **positive nutrient balance** would be accumulating a nutrient within the body. This would happen during periods of growth, for example when minerals were being deposited in bones. The minimum amount of a nutrient required to achieve nutrient balance is regarded as the **physiological requirement** for that nutrient.

1.2 Recommendations for nutrient intake

Although each person has his or her own physiological requirement for each nutrient, the main influences on requirements are age, sex, body weight, pregnancy and lactation. These factors are taken into consideration when compiling **RDA (Recommended Daily Amount** also termed **Dietary Reference Values** or **DRVs** in the United Kingdom) tables for humans (see Appendix).

Most countries publish RDA or DRV tables for various human groups. In addition, there are international tables published by the World Health Organisation. RDAs or DRVs are derived from data on physiological requirements obtained from deficiency and balance studies and are compiled by committees of experts who examine the world literature. All the various national and international committees have access to the same data, but it is the interpretation of that data which varies from committee to committee and results in some very different recommendations from one country to another.

Figure 1.2 shows the basis for establishing an RDA or DRV from physiological requirements for a single human group, for example adult women.

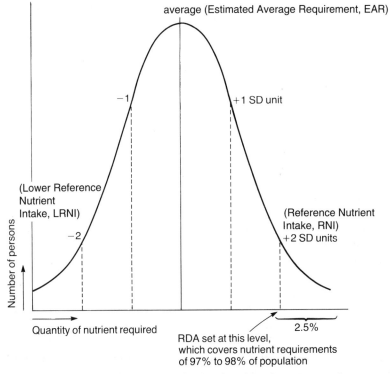

Figure 1.2 Normal distribution curve for the physiological requirement for a nutrient. The minimum RDA (Recommended Daily Amount) is the mean + 2 standard deviations (SD). Calculated in this way the RDA should provide for the requirements of 97.5% of the population (the standard deviation is a statistical measure of the variability of the data). Dietary Reference Values (DRVs) for the United Kingdom are in capitals.

In addition to the extra two standard deviations added to the mean, RDAs may also contain an extra amount as a 'safety margin'. For most nutrients this is sensible, since no harm would come from an intake of a nutrient somewhat higher than the requirement. However, this is not the case for energy intake, where intake should meet requirements and no more, since excess energy intake will lead to obesity which can be associated with ill health. Thus for energy, the RDA is set at the mean physiological requirement for a particular human group, without the addition of any safety factors. In the UK this is called the EAR or Estimated Average Requirement (see Figure 1.2).

RDAs or DRVs are designed for groups of people, and, although they are used as a yardstick to compare dietary intake of people from dietary surveys (see below), **they cannot be used to indicate that a particular person is malnourished.** The RDA or DRV of a nutrient is not the requirement for that nutrient of that individual, but the recommendation of that nutrient **for a group of people.** Because of the way it is derived it is likely to be higher than the physiological requirement. However, we do not know the physiological requirement of most individuals for nutrients as they have not been measured and, therefore, we have only the RDA or DRV to go by. If the intake of any particular nutrient for an individual is below the RDA or DRV then this may **indicate** that a person may be **'at risk'** of being undernourished, but to classify the person as malnourished would need clinical or biochemical evidence.

The amounts of vitamins recommended for various groups of humans in the UK are given also in the Appendix. As well as those mentioned in the Appendix there are others required (see Table 1.2) which have now been assigned either a DRV or a Safe Level by the Department of Health (1991).

1.3 Dietary surveys and composition of foods' tables

The best way to determine that a person is not getting adequate nutrients is to look for clinical signs of deficiency. This is, of course, rare in western societies for most nutrients, except perhaps iron. Nevertheless, marginal intakes, leading to sub-clinical malnutrition, may occur and these could be detected by appropriate biochemical tests on blood samples. Biochemical studies are costly and difficult to undertake, and, although it has already been mentioned that measuring the **intake** of nutrients cannot prove malnutrition, this information can be used to identify those who may be at risk of malnutrition. Intake data are fairly easily collected and can be used to determine trends over time and to compare the nutrition of one group with another.

Dietary surveys can be carried out in a number of ways. One way would be to determine the amount of food produced in a country, add the amount imported, subtract the amount exported and divide by the number of people in the population. This is called the **balance sheet method** and is used in strategic planning, but it is never accurate because so many assumptions are made. For example it is difficult to obtain accurate census figures.

Dietary surveys can be carried out on an institutional basis in a similar way. The National Food Survey in the UK is a survey carried out every year by the Ministry of Agriculture, Fisheries and Food since 1952, and is done on some 7000 households in various geographic locations. Food eaten in the household is recorded, along with data about those eating it. Data are converted to nutrients, using tables of food composition, so that the intake of the average person can be calculated.

The best type of dietary survey is the **weighed dietary survey**. This is carried out on an individual basis usually for seven days, during which time the individual weighs and records everything which he or she eats and drinks. From these data the nutrients eaten can be calculated from composition of foods' tables, an example of which is McCance & Widdowson's *The Composition of Foods* (1978) by Paul & Southgate. Because of the large number of different foods consumed each day the calculation would become very tedious, but for the help of the computer.

1.4 Dietary guidelines

In western countries, malnutrition is not normally encountered or expected, but some groups are at risk of nutrient deficiency according to many dietary surveys. As in most industrialised countries, however, overnutrition is now seen as a greater threat to public health than undernutrition. There is a whole range of diseases which are now considered to be influenced to some extent by dietary imbalance or excess, and these are often referred to as the '**diseases of affluence**' as they particularly affect the well-to-do (that is the general population of developed countries and the higher echelons of Third World countries). Of particular concern is the influence of diet on the development of coronary heart disease which accounts for a very large proportion of the deaths in the UK. The relationship between diet and coronary heart disease will be discussed in Chapter 8.

The last few years have seen the publication of two influential reports, the **NACNE** (1983) and **COMA** (1984) reports. The recommendations in these reports are now commonly known as the **dietary guidelines**, and a summary of them is given in Table 1.3. Nutritionists have long considered that diet and health are related, but these documents were the first to suggest actual levels of dietary components to aim for. The recommendations do not receive universal support from all nutritionists: some feel that the evidence linking diet to disease patterns is not strong enough to make recommendations at this stage. However, many are concerned that something should be done to inform the public, even if the message is incomplete and subject to alteration.

Table 1.3 Dietary guidelines from NACNE and COMA reports

Dietary component	NACNE (1983)	COMA (1984)
Total energy	maintain ideal weight	avoid obesity
Total fat	↓ 30% TE	↓ 35% TE
Saturated fat	↓ 10% TE	↓ 15% TE
Cholesterol	no recommendation	
Dietary fibre	↑ 30 g per day	↑
Starch (as whole-grain cerial)	↑	↑
Sugar	↓ 20 kg yr⁻¹	↓
Salt	↓	do not increase
Alcohol	↓ 4% TE	no recommendation

↑, increase to; ↓, decrease to; TE, energy supplied by component as % of total energy in the diet.

If the dietary guidelines are to be implemented in the UK, we will need to modify our diet to come more into line with that of a country of intermediate development. Typical figures for average diet composition in a rich country such as the UK for fat, starch and sugar (as energy from dietary component, as a percentage of total energy of the diet) are given in Figure 1.3 (also see Chapter 6), compared with that of a poorer country and one of intermediate development.

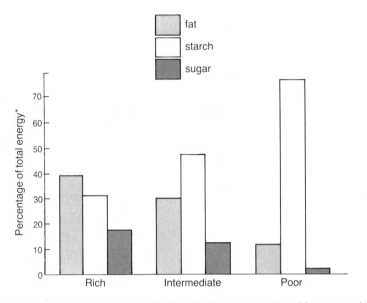

*To be correct, this proportion is the individual food calculated as an energy value and then expressed as a percentage of the total energy in the diet

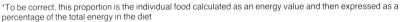

Figure 1.3 Contribution of major energy sources to dietary energy, according to country

The publication of the guidelines has led to a much greater interest in nutrition on the part of those working in the Food Industry. The guidelines pose an enormous challenge in terms of new product development, to provide foods to enable consumers to choose a 'healthy' diet.

2 Basic chemistry of major food substances

Carbohydrates, fats and proteins can all be used as an energy source by the body, although the major function of dietary protein is for synthesis of body proteins. These aspects are dealt with in Chapters 4 and 6, but a knowledge of the basic chemistry of these major food substances is essential for an understanding of the processes of digestion, absorption and metabolism (Chapters 3 and 4).

2.1 Carbohydrates

Carbohydrates are synthesised by plants during photosynthesis. This is the process of trapping energy from sunlight with the aid of chlorophyll. Carbon dioxide, taken in from the air, and water are combined using this energy into carbohydrate for use as food for the plant. The energy contained in the carbohydrate can then be released for use by the plant at some time in the future by being converted back to carbon dioxide and water (a process called respiration). Alternatively, the plant may be eaten by animals, including humans, and the energy in the carbohydrates used by them in a similar way.

There is a wide range of different carbohydrates in nature, the least complex of which are the simple sugars called **monosaccharides**. Glucose, fructose and galactose (see Figure 2.1) are the most common ones, and, although little is found free in foods (glucose and fructose are found in fruit juices), these provide the building blocks (monomers) for the synthesis of more complex carbohydrates, such as starch.

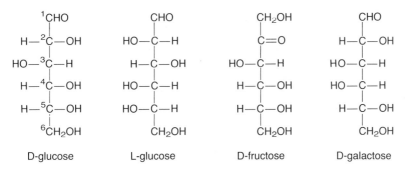

Figure 2.1 Structures of D- and L-glucose, D-fructose and D-galactose

Glucose is an optically active substance, as carbons 2, 3, 4 and 5 are asymmetrical. This means that each carbon has four different groups attached to it. A molecule with an asymmetric carbon atom will exist in another form which is the mirror image of itself. These are able to twist, in equal and opposite directions the plane of polarised light (light which vibrates in one plane only). In nature, glucose is dextrorotatory (turns the plane of polarised light to the right), and is designated as being 'd' or '+'. Laevorotatory glucose is unknown in nature, but it can be made in the laboratory.

In fact, glucose exists mainly in a cyclic form which is in equilibrium with the straight-chain form, and this is shown diagrammatically in Figure 2.2. When in the ring form, carbon 1 atom also becomes asymmetrical to give two possible structures, known as the alpha and beta forms of glucose which are shown also in Figure 2.2. These forms are interchangeable as there is an equilibrium mixture of the two forms, via the open-chain form.

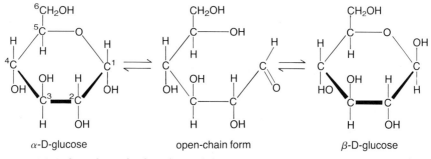

α-D-glucose   open-chain form   β-D-glucose

Figure 2.2 Cyclic and straight-chain forms of glucose

Two simple sugar molecules joined together are called a **disaccharide**. Disaccharides are found in larger quantities in foods than simple sugars. **Sucrose** (Figure 2.3), composed of glucose and fructose, is found in fruits, and **lactose**, composed of glucose and galactose, is uniquely found in milk. **Maltose**, composed of glucose only, is the end-product of the breakdown of starches by amylase which occurs during beer manufacture in the malting of barley and also in the digestion of starch in the human gut (Chapter 3).

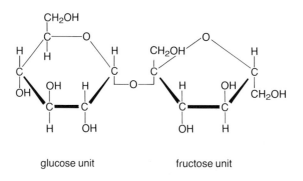

glucose unit   fructose unit

Figure 2.3 Structure of sucrose

Polysaccharide carbohydrates all have high relative molecular masses and are generally insoluble in water. They are synthesised in the plant from simple sugars. Although **starch** can be regarded as an energy store, other plant polysaccharides form part of the stucture of the cell wall and comprise **dietary fibre**, which is described in more detail in Chapter 6. Dietary fibre is composed of **cellulose**, **lignin** and **hemicellulose**. The first two are loosely defined chemical entities, but there is much variability in the composition of hemicellulose. In animals, the only polysaccharide of importance is **glycogen**, which is a carbo-hydrate store analogous to starch in plants.

Starch is found in many foods, including seeds and roots where it is present in granules. These vary in shape and size according to source. Large starch granules (such as in potato) are difficult to digest in the raw state as digestive juices cannot penetrate the crystalline structures within the granule. (The crystallinity is caused by the alignment of the starch molecules, leading to formation of hydrogen bonding between adjacent chains. For an example of hydrogen bonding in proteins, see Figure 2.11.) After cooking, digestion is facilitated by gelatinisation of the starch granule. Gelatinisation occurs when water penetrates the starch granule, causing the crystalline structures to break up and become hydrated. Water penetration is facilitated by heating or even by mechanical damage to the starch granule.

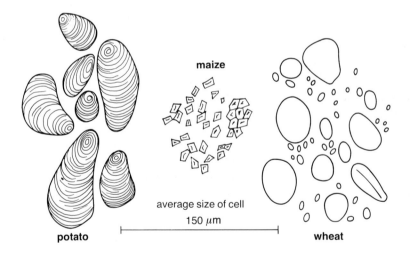

Figure 2.4 Examples of starch granules

Starch is composed only of the alpha forms of glucose units. Superficially it would seem to have a similar structure to cellulose which is also composed of glucose units. However, the glucose in cellulose is in the beta form. The difference in the structure of the two substances (Figure 2.5) is fundamental to their digestion and utilisation as an energy source in humans: starch is normally readily broken down by digestive enzymes which have no effect on cellulose.

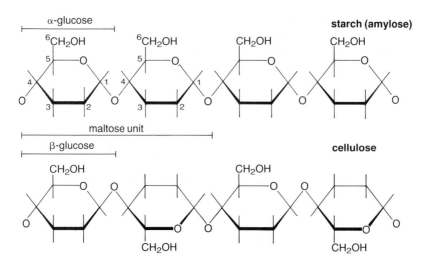

Figure 2.5 A comparison of the structures of starch and cellulose

Starch is a mixture of two polymers, **amylose** and **amylopectin**. Most foods contain much more amylopectin than amylose. Amylose has a lower relative molecular mass than amylopectin and is not branched (Figure 2.5) as is amylopectin. The branched structure renders amylopectin more difficult to digest than amylose.

Cellulose is the main structural carbohydrate of plants. It consists of thousands of glucose units, but cannot be digested by humans because an enzyme necessary to rupture the links between the beta forms of glucose (Figure 2.5) is not secreted in the gut. However some animals can utilise cellulose as an energy source as they have microorganisms in the gut which can break it down to glucose. These are found in the rumen (special stomach) of ruminant animals, such as cows, or in the adapted large intestines of herbivores, such as rabbits and horses.

Other plant polysaccharides are **hemicelluloses, pectins** (which are used to make jam set) and many natural **gums**, like the guar gum commercially extracted from the cluster bean grown in the tropics. In addition to being located in the cell wall, pectins are typically found in the middle lamella between plant cells and act to 'cement' the cells together. During cooking of plant foods, like potatoes, enough time has to be given to ensure that the pectins are broken down to give a 'cooked' texture.

Hemicelluloses may be composed of **hexoses** (monosaccharides with six carbons, such as glucose and fructose), **pentoses** (five carbons, such as arabinose) or other carbohydrate units called **uronic acids**. The presence and quantities of these will vary from one plant to another. Hemicelluloses are the most variable component of dietary fibre; their composition and nature in comparison to cellulose and lignin and their specific composition in a particular food give rise to the differences in physiological activity of different dietary fibres (Chapter 6).

Lignin is laid down in the plant as it gets older, as a strengthening material (secondary thickening). There is only a small amount in our food ($<2\%$); indeed, large amounts would be unpalatable and give food the texture of wood fibre. Lignin is not a carbohydrate, although often loosely classified as such. It has a polymeric structure based on phenylpropane monomers.

2.2 Proteins

As well as forming the structure of many foods, proteins are essential in the body for many functions. They can be structural, such as collagen (in bone or cartilage), or functional (enzymes or many specialised proteins like haemoglobin). Proteins in our food are broken down by digestion to their constituent **amino acids**. After absorption. these are reassembled to form body proteins.

There are about 20 amino acids commonly found in proteins. All the amino acids are alpha amino acids, that is the amino group, $-NH_2$, is linked to the alpha position (Figure 2.6).

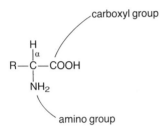

where the R group varies, depending on the amino acid

Figure 2.6 The generalised structure of an amino acid

Glycine is the simplest amino acid, where R is hydrogen, but other amino acids may be much more complex (Figure 2.7).

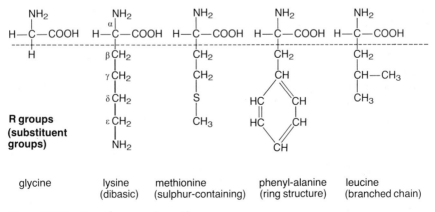

Figure 2.7 Structure of some amino acids

As the alpha carbon atom has four different groups attached to it in all amino acids except glycine, this means that it is asymmetrical and therefore potentially two optically active forms exist which are the mirror image of each other. These are known as the D and L structures. In nature, amino acids are almost exclusively in the L structure.

Amino acids are classified as neutral, acidic or basic. All amino acids contain one amino and one carboxylic group, as already indicated, and as such are called neutral. In addition, basic amino acids possess an extra basic group, and acidic amino acids have an extra carboxyl group.

One of the most outstanding properties of amino acids is that they are **amphoteric**. This means thay can act as an acid or a base. When dissolved in water amino acids ionise, as in Figure 2.8. This form is called a **zwitterion** or

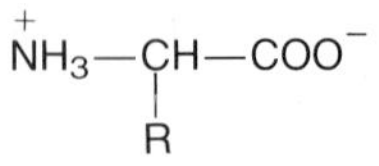

$$\overset{+}{N}H_3 - \underset{\underset{R}{|}}{C}H - COO^-$$

Figure 2.8 Ionisation of an amino acid in water

dipolar ion as it carries both positive and negative charges, and the extent of ionisation depends on the pH of the solution (Figure 2.9).

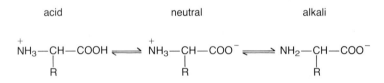

Figure 2.9 Ionisation of an amino acid at different pHs

The amphoteric nature of amino acids means that they can act as buffers. Up to a point, amino acids (and proteins) are capable of combining with acids or bases to prevent change in pH. This is very important in biological systems to maintain **homeostasis** (the consistent quality of the internal environment of the body).

The pH at which opposite charges on the zwitterion are equal (that is, net charge is zero) is called the **isoelectric point**. The isoelectric points for proteins are used by food manufactures to produce protein isolates (such as from soyabean). Protein is removed from the seed by solubilisation at a pH away from the isoelectric point, and then collected by precipitation, by changing the pH to the isoelectric point. Acidic and basic amino acids have different isoelectric points from neutral amino acids ($<$pH5, $>$pH7 and pH7 respectively).

Two amino acids can combine to form a **dipeptide** by means of the **peptide link**. The reaction is shown in Figure 2.10. A dipeptide may go on to react with

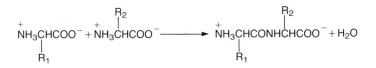

Figure 2.10 The formation of the peptide bond

another amino acid to produce a **tripeptide** in a similar manner, and this is the basis of protein synthesis. In nature, it is possible for amino acids to be added in this way time after time to build up a protein. This is called the **primary structure** of a protein. Even simple proteins may have a relative molecular mass (molecular weight) of many thousands. Protein synthesis in all cells of the body is directed by **DNA** (deoxyribose nucleic acid, the genetic material in the nucleus of the cell) via **RNA** (ribose nucleic acid).

The structure of proteins is like a string of beads, with R groups projecting from the central axis. In a protein, the chain is not usually extended but twisted in a **secondary structure**, a common form is then further coiled like a spring, called an alpha helix (see Figure 2.11). Secondary structures are stabilised by **hydrogen bonding** between certain groups in the chain. The **tertiary structure** of a protein determines its molecular shape (that is, globular or fibrous molecules) and usually includes sections in the alpha helix form. It is maintained by various types of bonding, including covalent bonds.

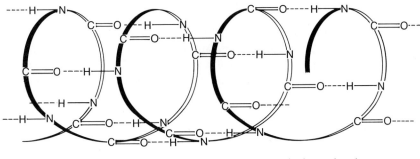

----- hydrogen bond

Figure 2.11 The secondary structure of proteins

Protein quality

Protein of 'high quality' has high digestibility and an amino acid pattern whose level of essential amino acids (those required in the diet) closely reflects human amino acid requirements. In diets where one of the essential amino acids is present at a much lower level than that required, this becomes the **limiting**

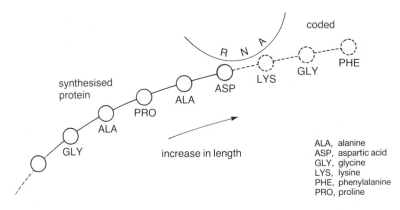

Figure 2.12 Protein synthesis on a lysine-deficient diet

amino acid because its deficiency limits the use of the other amino acids. This is best explained by the example given in Figure 2.12. This shows the synthesis of a single protein in the body when a person is eating a diet deficient in lysine. Lysine is an essential amino acid which is low in most cereals, including wheat. (Of course, within the body many different proteins are being manufactured simultaneously, but the point can best be illustrated by thinking about the manufacture of a single protein.)

As Figure 2.12 shows, proteins are assembled from amino acids in sequence according to codes on the RNA. If you imagine this protein being assembled, all will go well until there is a requirement for lysine. At that point none is available, synthesis will stop and other amino acids coded to come after lysine cannot be utilised (glycine and phenylalanine in this case). Therefore, the use of glycine and phenylalanine has been limited by the low level of lysine. As many proteins are assembled at the same time, there will come a point when a low level of lysine will inhibit the use of all the other amino acids present. This is what is meant by the concept of the limiting amino acid; the absence of one essential amino acid prohibits the use of the rest.

Protein complementation

As cereals are such a large part of most people's diet, it might naturally be thought that their low level of lysine may prejudice good nutrition. In practice this does not usually occur for two main reasons.

(1) The amounts of protein eaten are normally much in excess of requirements and normally the limiting amino acid is present at a low level and not absent completely. Therefore, eating more protein than is required means that more of the limiting amino acids become available to the body.

(2) Single foods are rarely eaten as the sole item of the diet. Cereals in the average UK diet are largely supplemented by meat and animal products which are high in all the essential amino acids, including lysine. Proteins can complement one another, so that if one is high in lysine but low in another essential amino acid, and a second protein is low in lysine but high in all other amino acids, the protein quality of the two together is better than each individually. Combinations of proteins which show these characteristics are cereal/legume mixes, eaten a lot in the Third World, where the high lysine content of legumes complements the low lysine content of cereals, while the higher content of the sulphur-containing amino acids (methionine and cystine) in cereals complements the low level in the legumes.

As most people consume at least double the amount of protein they require per day, the only times when the limiting amino acid content of a diet may become a problem is in very poor areas of the world, where the staple in the diet is exceptionally low in protein or protein quality and the energy intake is lower than requirements (see Chapter 8).

2.3 Fats

Fats contain **essential fatty acids** which are necessary in our diets for the formation and integrity of cell membranes. In addition, fat adds palatability to our diets, a very important factor in poor countries where foods may be monotonous. Oils and fats are made up from fatty acids and belong to a larger group called **lipids**, which are generally insoluble in water. Other lipids important in the body are **phospholipids**, which also contain fatty acids, **cholesterol** and related compounds (bile salts) which are not composed of fatty acids.

Edible oils and fats are composed of **glycerol** and fatty acids. They used to be called triglycerides, but this older term has been more recently replaced by the term **triacylglycerols**. The structures of glycerol, a triacylglycerol, a **monoacylglycerol** and a **diacylglycerol** are seen in Figure 2.13. Monoacylglycerols and diacylglycerols are formed during digestion.

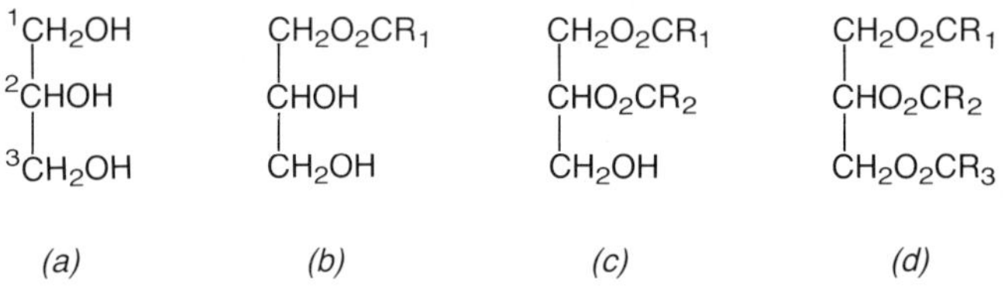

$$^1CH_2OH \qquad CH_2O_2CR_1 \qquad CH_2O_2CR_1 \qquad CH_2O_2CR_1$$
$$^2CHOH \qquad CHOH \qquad CHO_2CR_2 \qquad CHO_2CR_2$$
$$^3CH_2OH \qquad CH_2OH \qquad CH_2OH \qquad CH_2O_2CR_3$$

(a)   (b)   (c)   (d)

Figure 2.13 Structure of (*a*) glycerol, (*b*) mono-, (*c*) di- and (*d*) triacylglycerol

There are a number of different fatty acids which are common in our diets. The structures of the four most common are given in Figure 2.14.

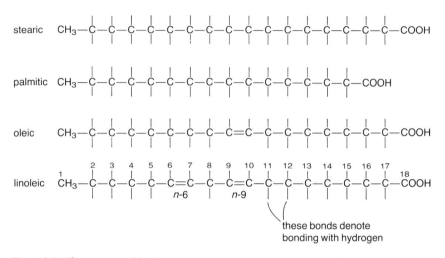

Figure 2.14 The structure of the four most common fatty acids in our diet. (Note the numbering of the carbon atoms allows us to denote unsaturated groups as *n*-6, *n*-9 etc.)

In Figure 2.14 we can see examples of three different classes of fatty acids, so classified because they have different physical properties and different physiological functions in the body. These are **saturated**, **monounsaturated** and **polyunsaturated** (PUFA) fatty acids. (Unsaturated means that some carbons are joined to others by a **double bond** rather than a single one. The double bond has the potential of being partially broken to accept two more hydrogen atoms, thus becoming saturated.) Some PUFA are essential nutrients (Chapter 6) which are necessary for cell membrane structure. Natural fats are mixtures of different fatty acids, but the greater the amounts of saturated fatty acids they contain the harder (more solid) they are. In contrast, oils, such as sunflower seed oil, contain a high proportion of PUFA.

Hard fats are mainly, but not exclusively, of animal origin and contain more saturated fatty acids, while oils and soft fats contain more PUFA. Food manufacturers are able to produce hard fats from oils by hydrogenation (in this process some of the unsaturated bonds are saturated), so that the products can be used in margarine. These days there is a swing away from hard margarines and saturated fats, towards eating more PUFA for health reasons (Chapters 1 and 8).

Rancidity is the spoilage (development of 'off' flavours) in fats or oils on storage. Edible oils are particularly susceptible to oxidative rancidity, especially if they have been refined and the natural **antioxidants** (such as vitamin E) have been removed. Fish oils contain much less vitamin E than oils from plants and, therefore, deteriorate very rapidly indeed.

Rancidity can occur by two mechanisms: **hydrolytic** and **oxidative rancidity**. The former occurs when the triacylglycerol molecule breaks up to glycerol and free fatty acids on contact with water in the presence of enzymes or contaminating microorganisms. (Rancidity of butter is due to this: it releases the free fatty acid, butyric acid, which has a very strong flavour.)

Oxidative rancidity occurs as a result of the reaction between oxygen from the atmosphere and unsaturated fatty acids. Some end-products of this reaction are **aldehydes** and **ketones** which have unpleasant flavours. Once started, this form of rancidity is self-propagating in a chain reaction. It can be started by the production of free radicals; these are very reactive molecules which can develop in an oil or fat in response to oxygen exposure and can be promoted by the presence of traces of metals. The formation of free radicals is also promoted by heat and light. The free radicals will react with double bonds, eventually breaking them, releasing the aldehydes and ketones. It is possible to stop the reaction by removing the free radicals, which is the function of antioxidants.

3 Digestion: alimentary canal, digestion and absorption

3.1 The necessity for digestion

Food needs to be digested so that relatively simple compounds are released from complex polymers for two main reasons. These are so that the body can absorb and utilise them. The fact that food is taken into the mouth and swallowed does not mean that it enters the body in the physiological sense. From the physiological point of view, food in the lumen of the gut is still regarded as 'outside' the body. Only after absorption through the intestinal **mucosa** is the food truly regarded as part of the body. Mucosal cells play a unique and essential role in the body in that they absorb nutrients not only for their own metabolism, but also for the metabolism of the body as a whole. To some extent the mucosal cells can also protect the body by not absorbing some harmful substances or by regulating absorption to avoid excess entering the body.

Food digestion is also required to provide the 'building blocks' for synthesis of the body's own proteins. Thus, food proteins need to be broken down to their constituent amino acids so that the body can use the individual amino acids to form its own unique proteins, determined according to genetic information. Complex (polymerised) carbohydrate, such as starch, needs to be broken down to simple sugars, either for use as an immediate energy source or to be stored as glycogen or fat for future use as an energy source. Similarly, dietary fats are broken down during digestion and reassembled in typical 'human' form for structural use or as a store for use as an energy source.

3.2 The structure of the alimentary canal

Figure 3.1 shows the structure of the alimentary canal in humans. Digestion commences in the mouth, where the food is chewed by the teeth into small pieces which make it easier for subsequent digestion. In the mouth food is also mixed with saliva which acts as a lubricant to form the food into a **bolus** (ball) and to facilitate swallowing. Saliva also contains an enzyme, **salivary amylase**, which can break down starches to maltose (see overleaf).

A generalised diagram through the gut wall is shown in Figure 3.2. The main features in Figure 3.2 are present throughout the gut, although with modifications depending on the position. The action of the powerful musculature under the outer serosal coat of the wall is important in digestion for mixing the gut contents and for propelling food through the tract. In the upper part of the tract food is moved onward by **peristalsis** (the contraction of the muscles behind the food bolus and the relaxation of the muscles in front). However, in the large intestine food is propelled forward by random segmentation of the gut wall caused by muscular contraction.

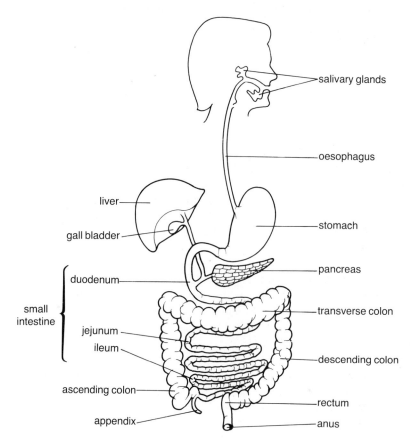

Figure 3.1 The structure of the human alimentary canal

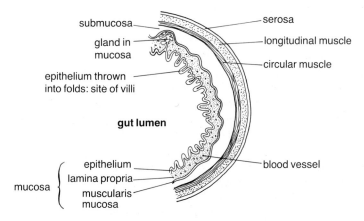

Figure 3.2 Generalised diagram of the gut wall

3.3 Structure of the mucosa of the small intestine

The gastrointestinal tract is 380 cm long, but the very large surface area of the mucosa is due to it being formed into villi (see Figure 3.3). This gives a surface area of about 300 square metres. In addition, the **brush border** (microvilli on the surface of the mucosal cells) further increase the area. At the base of the villi are **crypts of Lieberkuhn**, where the cells of the mucosa are formed. These cells migrate to the tips of the villi and have a lifespan of 1–3 days only. Inside each villus is a capillary network and a lacteal. The capillaries drain into venules and then to the portal vein. The portal vein carries absorbed end-products of digestion from the gut to the liver. The lacteals or lymphatic capillaries drain into larger lymphatic vessels and finally into the bloodstream via the left thoracic duct in the neck region. The lymphatic system is a complementary system to the blood circulatory system, with some separate functions.

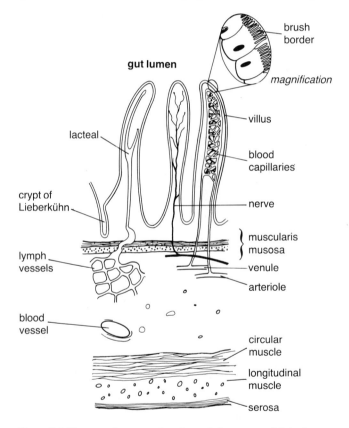

Figure 3.3 Diagram of cross-section through human small intestine

3.4 Control of gut secretions

Proper control of gut secretions is very important to prevent digestive juices being unnecessarily produced when no food is eaten, and to ensure that adequate secretions are present to complete digestion after a large meal. Control of **gut motility** (movement) and the **secretion of gut juices** is partly under nervous control, through the **vagus** nerve from the brain. It is because of this nervous control that the emotional state of the person will affect gut activity. As well as nervous control, both gut motility and secretion are under **hormonal control** by **gastrointestinal hormones.**

We now know that there are a large number of hormones that regulate aspects of gut function and most are peptides composed of amino acids. Research is very active in this area at the moment and many new discoveries are still being made about the regulatory processes involved in gut function. For a summary of the most important ones see Table 3.1.

Table 3.1 Action of the major gastrointestinal hormones

Hormone	Site of secretion	Action
Cholecysto-kinin (CCK)	all small intestine	Gall bladder contraction ↑ pancreatic enzyme secretion ↑
Gastric inhibitory peptide (GIP)	duodenum jejunum	Stomach acid secretion ↓ insulin release from pancreas ↑ intestinal mucosa secretion ↑
Gastrin	stomach duodenum	HCl release by stomach ↑ pancreatic enzyme secretion ↑ motility of intestine ↑
Secretin	duodenum jejunum	Pancreatic bicarbonate and water ↑ gastric acid secretion ↓ motility of intestine ↓

Adapted from Pike & Brown (1984). ↑, stimulates; ↓, inhibits.

One of the most interesting features is that **cholecystokinin** (CCK) appears to affect the regulation of feeding behaviour in addition to the functions described in Table 3.1. It is known that food intake in animals is reduced when CCK is introduced into brain cells. Another hormone which is now known to have a more wide-ranging function than previously thought is **gastrin**. The injection of gastrin into the **hypothalamus** (part of the forebrain responsible for regulating food intake) increases the secretion of gastric acid in rats. Indeed, gastrin has now been shown to stimulate RNA, DNA and protein synthesis along almost the entire length of the gut and in this way can increase the amount of tissue in the gut; probably important when people are eating larger quantities of food.

3.5 Digestion: carbohydrates

The only polysaccharides which are digested to any degree by humans are starches. Other polysaccharides in the human diet include celluloses and hemicelluloses. Although these cannot be broken down by the digestive enzymes, some are fermented by the microflora of the large intestine and the short-chain fatty acids so produced may be absorbed by the large intestine and used by the body as an energy source.

Starches are broken down to start with in the mouth by **salivary amylase**. This breaks down cooked starch to maltose, but its activity depends on the time the food is in contact with the enzyme before the pH is reduced in the stomach. The optimal pH for this amylase is about 6.0–6.6 and, of course, in the stomach the pH is much lower (1–2). However, in the fundus (first part) of the stomach, it takes some time for the pH to lower and during this time the amylase is active. The amount of breakdown by this enzyme may be quite considerable.

The major enzyme for breaking down starches is amylase from the pancreatic juice, which can act on cooked or raw starch. It hydrolyses $\alpha 1,4$ glucosidic bonds (Figure 2.5), giving maltose as the end-product. If branched starches (amylopectin) are present then dextrins of various molecular weights are also formed. No further hydrolysis of carbohydrates takes place in the intestinal lumen since the **disaccharidases** are located in the mucosal cells. Indeed, hydrolysis of disaccharides on the external surface of the mucosal cell (membrane digestion) is midway between digestion and absorption and ensures absorption at the final stages of hydrolysis. The main disaccharidases are **sucrase** (end-products being fructose and glucose from sucrose), **maltase** (glucose from maltose) and **lactase** (glucose and galactose from lactose).

3.6 Digestion: lipids

The presence of fat in the duodenum is said to cause the release of the hormone **enterogastrone**, which decreases gastric (stomach) secretion and mobility and thereby slows gastric emptying time. However, the exact mechanism operating (and indeed the existence of enterogastrone) is still open to question. Nevertheless the rate at which fat enters the duodenum from the stomach is regulated by, and appears to be correlated with, the capacity of the enzymes from the pancreas to digest fat entering the duodenum from the stomach.

The major digestive **lipase, glycerol ester hydrolase**, is present in pancreatic juice, and hydrolyses (breaks bond) glycerol esters (at the 1 and 3 position – see Figure 2.13a). This enzyme needs bile salts for its action. **Bile salts** are present in bile which is produced by the liver and stored in the **gall bladder**. The gall bladder contracts during a meal to release the bile into the small intestine. Bile salts, fatty acids and glycerol all have a detergent action on fat and aid in the **emulsification** of triacylglycerols.

End-products of lipid digestion are **monoacylglycerols** and **fatty acids** which, with the bile salts, form **micelles** (particles of fatty material with a thin coating of bile salts, which enables them to remain stable in the water-based secretions of the gut). These micelles combine with (dissolve) free fatty acids, cholesterol and fat-soluble vitamins and are then brought into contact with the mucosa. The mucosal cells contain lipase, which continue digestion and aid absorption of lipid material. Micelles are very important in lipid absorption. This is the way the fat-soluble components of the diet are carried in the aqueous phase of the gut lumen. However, exactly how the material in the mixed micelle is taken up by the mucosal cell is not completely understood. It is well known that bile salts are reabsorbed in the ileum, pass along the hepatic portal vein and are reused by the liver for bile formation, a cycle called the **enterohepatic circulation**.

3.7 Digestion: proteins

The chemical and mechanical stimulation produced by the presence of food in the stomach causes the release of the hormone **gastrin**, which in turn stimulates the **parietal cells** in the stomach wall to produce **hydrochloric acid**. Hydrochloric acid has a very important antiseptic function. (Note that although the contents of the small intestine in humans are nearly sterile under normal conditions, in some animals, such as the rat, this is not the case.) Other functions of the hydrochloric acid are to swell protein, which aids digestion, and to activate pepsinogen (**precursor** of **pepsin**). Once pepsin is formed it can activate more pepsinogen.

Pepsin hydrolyses the peptide bonds that link the amino acids tyrosine and phenylalanine to other amino acids and gives polypeptides of varying lengths as end-products of the reaction. The action of pepsin stops when the stomach contents are released into the duodenum and the pH is raised. The action of the acid **chyme** (the liquid food mix) entering the duodenum causes the secretion of the hormone **secretin**. This hormone stimulates the secretion of pancreatic juice rich in **bicarbonate** but poor in enzymes, which tends to neutralise the acidity of the chyme. The products of protein digestion entering the duodenum stimulate the secretion of another hormone, **cholecystokinin (CCK)**, which increases the enzyme content of the pancreatic juice.

Pancreatic juice contains enzyme precursors **trypsinogen, chymotrypsinogen** and **procarboxypeptidase**. An enzyme secreted from the intestinal mucosa, called **enteropeptidase**, converts trypsinogen into **trypsin** by removing a peptide from the molecule, and once the trypsin is formed it will itself activate all the other pancreatic enzyme precursors, including trypsinogen. Activation is achieved in each case by breaking a peptide bond next to an arginine or lysine residue close to the beginning of the protein chain.

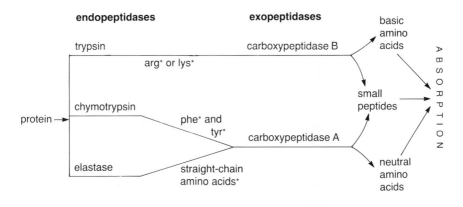

*amino acids at the carboxyl end of the peptide chain;
arg, arginine; lys, lysine; phe, phenylalanine; tyr, tyrosine

Figure 3.4 Action of endopeptidases and exopeptidases from the pancreas on protein digestion

There are two types of pancreatic **proteases** which work together to break down proteins (see Figure 3.4). One type is the **endopeptidases**, such as trypsin, **chymotrypsin** and **elastase**, which break specific peptide bonds within the protein. Thus trypsin acts at a peptide bond between lysine, arginine or histidine and other amino acids and chymotrypsin breaks peptide links between tyrosine or phenylalanine and another amino acid. The other type of pancreatic protease comprise the **exopeptidases**, such as **carboxypeptidases A** and **B**, which break down the terminal peptide bonds at the carboxyl end of the peptide. A breaks down neutral or aromatic residues, while B breaks down basic residues (see Section 2.2).

The end-product of pancreatic enzyme digestion is a mixture of amino acids and short peptides. The final breakdown of peptides of from four to eight amino acids occurs in the mucosa. Indeed, the final stages of the breakdown of all short peptides occurs by the combined digestion/absorption activity of the intestinal mucosa (described in the next Section).

3.8 Absorption

Absorption of most nutrients takes place in the duodenum and the first part of the jejunum. Vitamin B_{12} is unusual in not being absorbed until it reaches the ileum.

Carbohydrate absorption

Of the monosaccharides commonly present in the diet or produced as a result of digestion, galactose is absorbed quickest, then glucose, followed by fructose. Glucose absorption is very efficient and nearly all the digested carbohydrate has been absorbed by the time the chyme reaches the lower portion of the jejunum.

Lipid absorption

Fat absorption occurs in two ways: some is absorbed into the blood vessels in the small intestine and then goes via the hepatic portal vein direct to the liver, and some is absorbed into the lymph system.

Via hepatic portal vein: This process only occurs for some medium-chain free fatty acids and monoacylglycerols. Free fatty acids with a chain length of 10–12 carbons or less are transported in free fatty acid form and leave the mucosa via the portal vein and are transported directly to the liver.

Via lymph system: Fatty acids and monoacylglycerols taken up by the mucosal cells are resynthesised to form triacylglycerols. These are formed into spherical **chylomicrons** by the addition of phospholipids, cholesterol, cholesterol esters and a specific protein, **apoprotein B**. In this way water-insoluble triacylglycerols can be transported as lipoprotein in an aqueous medium in the **lymph system**. Thus, lipids which have been hydrolysed, resynthesised and protein-coated enter the circulation via the lymph system as chylomicrons.

Protein absorption

The absorption of **intact protein molecules** may produce an **allergic reaction** in sensitive people. In the newborn, however, the gut is freely permeable to proteins and it is during the first two weeks of life that the **colostrum** (first milk after birth) from the mother carries such a high level of antibodies (gamma globulins) which are absorbed by the infant intact and give it some defence against infection. At this time of life other proteins such as egg protein if fed would also be absorbed intact and may provoke allergic response later in life. It is recommended to feed the very young on human milk during at least the first two weeks of life and preferably longer for this reason.

The end-products of protein digestion are free amino acids and small peptides. Amino acid absorption is rapid, but contrary to popular belief, di- and tripeptides have now been shown to be the main end-products of protein digestion rather than free amino acids. It is most likely that there is a single **carrier** system for all dipeptides in human mucosal cells. All protein digestion products leave the mucosal cell and enter the **hepatic portal vein** almost entirely as free amino acids.

Vitamin absorption

On the whole, and in contrast to minerals, vitamins are easily absorbed in the body, but there are a few cases where absorption is poor (low **bioavailability**). Some bound forms of **nicotinic acid** (niacin) occur in maize, called **niacytin**, which can only be released by the action of alkaline treatment. This is carried out traditionally in Mexico by the addition of limewater during food preparation. In other parts of the world, such as central Africa where a lot of maize is eaten and alkali treatment is not carried out, it is quite common to find cases of **pellagra**, the deficiency disease caused by a lack of nicotinic acid in the diet.

With water-soluble vitamins (see Table 1.2) there is very little problem of absorption in general, except with vitamin B_{12} which requires **intrinsic factor** which is secreted in the stomach. This is a **glycoprotein** (protein:carbohydrate ratio of 3:2), with a relative molecular mass of 50 000 daltons. The specificity of B_{12} and intrinsic factor is similar to the binding of antigen–antibody. After binding, the combined molecule is immune to digestion and proceeds to the ileum where there are receptor sites for absorption.

Fat-soluble vitamins (see Table 1.2) are, as the name suggests, associated in food with fat or lipid material. Poor absorption of fat can lead, in particular, to vitamin A deficiency, seen all too often in developing countries as permanent blindness with childhood being the dominant age group affected.

Mineral absorption

Many minerals are poorly absorbed by the body. To some extent this may be a protective mechanism to prevent the body becoming poisoned by excess intake. It is now known that many minerals are absorbed by special mechanisms, often a protein present in the mucosal cells. With some minerals, poor absorption is of little consequence as the diet contains much more than is required. But, iron and calcium provide good examples for which poor absorption can lead to deficiency conditions. However, if body supplies are low, or the blood level is low in a particular mineral, then absorption often increases in response.

Two main factors influence iron absorption, the state of iron stores and the activity of the bone marrow in red blood cell formation, as iron is needed for the synthesis of haemoglobin of red blood cells. The nature of the signals to the intestine is unknown, but as the level of iron in the diet is increased, the amount absorbed goes down. However, if a person is iron-deficient, then there is an increase in the amount of iron absorbed. Haemorrhage (including menstruation), growth and pregnancy all increase iron absorption. Thus women absorb more iron than men, average for men being 5.5% absorption, and for women 13% from a diet containing 10 mg day^{-1}.

The amount of dietary calcium absorbed is 50–70% in infancy (from milk), but only 10–40% in adulthood from a mixed diet.

3.9 Dietary factors inhibiting iron and calcium absorption

Minerals such as iron and calcium have low bioavailability, often because they combine with a number of components of the diet which render them unavailable.

Phytic acid can lead to the formation of insoluble iron salts. Phytic acid is present in wholegrain cereals. It has been known for some time that calcium is less available in wholemeal bread, even though it is present in higher amounts. The action of yeast in breadmaking produces phytase which splits phytic acid. One mystery is why those eating high oat diets such as those in parts of Scotland do not show increased rickets, as oats are high in phytic acid, and contain little phytase. It has been suggested that there is adaptation of the body leading to the presence of phytase within the gut.

Phosphates in excess can lead to the formation of insoluble iron salts.

Long-chain fatty acids can form insoluble soaps with calcium. Increased fat in the faeces (**steatorrhoea**) leads to increased calcium passed out of the body. Patients with steatorrhea may therefore develop **rickets** or **osteomalacia** after a time.

Oxalic acid combines with calcium to form insoluble calcium oxalate.

3.10 Factors increasing iron and calcium absorption

The presence of vitamin C encourages the formation of ferrous iron from ferric iron. Ferrous iron is more easily absorbed. Alcohol also increases iron absorption. Some wine drinkers in France have been found to absorb so much iron (red wine has a high iron content of 5–22 mg dm^{-3}) that it has lead to abnormal iron deposits in the body.

Vitamin D is necessary for the absorption of calcium, indeed most cases of calcium deficiency such as rickets are primarily due to a lack of vitamin D rather than of calcium. Vitamin D is produced by the action of sunlight on the skin, and it is important for the formation of a **calcium-binding protein** within the mucosa which is necessary for the absorption of calcium. **Protein** facilitates calcium absorption. It may be that the amino acids liberated in protein digestion form very soluble calcium complexes which are easily absorbed.

4 Metabolism of carbohydrate, protein and fat

4.1 Introduction

The end-products of digestion of carbohydrates, fats and proteins, as well as alcohol, can all be used by the body as energy sources. Energy release from these components does not occur in a single step, but in a controlled manner via **enzymic pathways** which are a series of enzyme reactions.

Breakdown of larger molecules into small ones is called **catabolism**. The catabolic products of glucose and fat metabolism are carbon dioxide gas and water. In the case of amino acids, urea is also a catabolic product. As well as being used as an energy source, end-products of digestion may be built up in the body to more complex structures, a process called **anabolism**. These products would be used for structural and functional purposes within the cell, depending on requirements. Catabolic reactions release energy which can be utilised by the body, while anabolic processes require an input of energy from the body.

4.2 Carbohydrate metabolism

After absorption, the end-products of carbohydrate digestion (that is the monosaccharide sugars galactose, fructose and glucose) are brought, via the bloodstream, from the intestine to the liver along the hepatic portal vein. Galactose and fructose are readily converted into glucose by the action of specific enzymes. Glucose can undergo a number of possible reactions, depending on the state of nutrition of the body. It can be synthesised into glycogen, oxidised to carbon dioxide and water, converted to fat or formed into amino acids.

The overall breakdown of glucose to release energy can be written as the **respiratory equation**. In this equation, each gram molecule of glucose (180 g) yields 2881 kilojoules (686 kcal) of energy.

$$C_6H_{12}O_6 + 6O_2 \rightarrow 6CO_2 + 6H_2O + \text{Energy}$$
$$180\,g \qquad\qquad\qquad 2881\,kJ$$

Of course, this equation represents a very crude impression of the way in which food (as glucose) is broken down by the body to release energy. In the body, this breakdown occurs in controlled stages called metabolic pathways. The energy released by these pathways can be trapped in **high energy phosphate bonds** in chemical substances such as **adenosine triphosphate** (**ATP**). Energy in these bonds is readily available, whenever needed, for chemical reactions.

Glycogenesis

If glucose is not immediately required as an energy source by the body then it is converted to **glycogen** by the process known as glycogenesis. Glycogen is a polysaccharide, similar in some respects to starch (see Section 2.1). Although it is formed in all tissues of the body, most is produced in liver and muscle where a certain amount can be stored until required. Glycogen can be readily converted back to glucose when required by the enzyme phosphorylase, a process known as **glycogenolysis.**

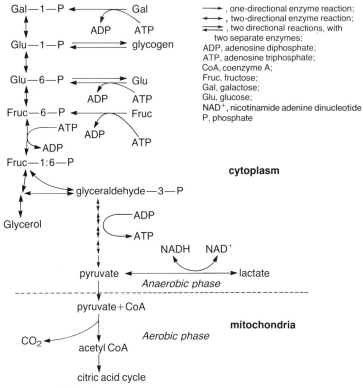

Figure 4.1 Glycolysis

Glycolysis

This is the oxidation of glucose to pyruvate by the **glycolytic pathway** (also called the Embden–Meyerhoff pathway), which is a series of enzymic reactions. Glycolysis occurs in every single cell of the body, in the cell cytoplasm. During this process, glucose (containing six carbon atoms) is broken down in steps to form **pyruvate** (a three-carbon compound). In glycolysis, the overall effect is to generate ATP (adenosine triphosphate) even though at some stages an input of ATP is required; the amounts produced compensate for this. The main steps of glycolysis are outlined in Figure 4.1.

Fate of pyruvate

Most of the pyruvate produced as a result of glycolysis undergoes further oxidation in mitochondria in the **citric acid cycle** (also known as the tricarboxylic acid or Krebs cycle which is described in the next section), although it can be converted into **lactic acid** (lactate). The conversion of pyruvate to lactic acid occurs in skeletal muscle undergoing work at such a rate that there is a lack of oxygen (called the oxygen 'debt'). In the presence of oxygen, the lactate is reconverted to pyruvate in the liver and the hydrogen formed in this reaction is attached to a hydrogen carrier, **NAD$^+$** (nicotinamide adenine dinucleotide) as NADH. NADH releases hydrogen through a series of reactions in the **respiratory chain** (see Figure 4.3) to oxygen, with the generation of ATP.

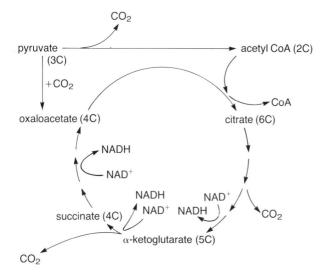

Figure 4.2 Citric acid cycle

The citric acid cycle

Pyruvate is capable of passing from the cytoplasm of the cell, through the **mitochondrial membrane**. Pyruvate oxidation, firstly by the citric acid cycle and then by the respiratory chain, occurs by a series of enzymes arranged in the correct sequence along the inner walls of the mitochondria. (Because of their importance in the release of energy from food, the mitochondria are sometimes called the 'power houses' of the cell.) Once inside the mitochondrion, two reactions of pyruvate are possible: the formation of oxaloacetic acid or acetyl coenzyme A (**acetyl CoA**), the latter through an irreversible process which produces CO_2 and ATP.

The oxaloacetic acid and acetyl CoA react together to form free CoA and citrate, and the cycle proceeds in sequence as shown in Figure 4.2. It proceeds from a 6-carbon molecule (citrate), through a 5-carbon to a 4-carbon molecule (oxaloacetate). Each time the cycle is completed, one molecule of acetyl CoA is used up and two molecules of CO_2 are liberated. In addition, eight atoms of hydrogen are 'released' and trapped as reduced NAD$^+$ or, as a related compound, reduced NADP$^+$ (nicotinamide adenine dinucleotide phosphate).

The respiratory chain

The electrons from the hydrogen trapped into reduced NAD$^+$ (or reduced NADP$^+$) are passed along the respiratory chain finally to oxygen. At the same time the energy is stored as ATP and regenerated oxidised NAD^+ or $NADP^+$ will again be available as a hydrogen acceptor. The regeneration of oxidised NAD^+ or $NADP^+$ occurs via a series of enzyme carriers, such as flavin and the cytochromes, which transfer electrons and hydrogen ultimately to oxygen

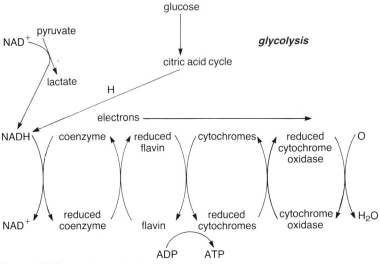

Figure 4.3 The respiratory chain

(Figure 4.3). In the absence of oxygen, of course, the respiratory chain will not proceed. Cytochrome oxidase is the final link in the chain, and it is this enzyme which will combine with cyanide (reversibly if the level can be reduced in time by the body's detoxication system). Some foods contain cyanogens (cyanide-producing substances) at sufficient levels to cause a public health problem (Chapter 9).

4.3 Protein metabolism

The end-products of protein digestion are amino acids (Section 3.7). Amino acids are the building units of all body proteins and some have special functions in addition. For example, arginine is part of the urea cycle (see below), and tyrosine is needed for the formation of the thyroid hormones that regulate the rate of metabolism.

Amino acids which are not used for the purposes mentioned above are broken down. The amino (NH_2) group is released as ammonia, which is converted to **urea** and excreted, or used for the synthesis of non-essential amino acids. The **carbon skeleton** that remains is used as an energy source or converted to glucose, fat or other amino acids. Thus the metabolism of carbohydrate, fat and protein is intimately linked in the body. This is reflected in Figure 4.4, which is a gross simplification of the breakdown of absorbed end-products of foods for the purpose of providing the body with energy. As well as showing the catabolic pathways for glucose (as already described), it also shows how end-products of fat and protein digestion (present in the body mainly as fatty acids and amino acids) are broken down and how they relate to glucose catabolism.

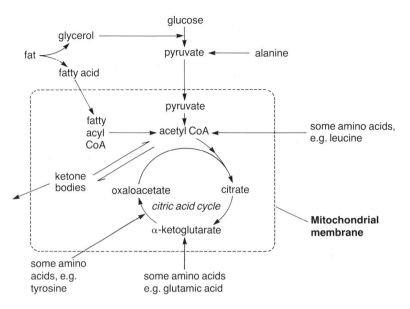

Figure 4.4 The interrelationships of fat, protein and carbohydrate catabolism

Deamination

The breakdown of an amino acid to the carbon skeleton (keto acid) with the release of the amino group is called deamination. For example, the amino acid alanine is broken down to the keto acid pyruvate, which can be broken down as an energy source as previously explained (see Figure 4.1).

Transamination

In some instances when an amino acid is deaminated, at the same time there is amination of a keto acid. This process is called transamination, and it is the process whereby non-essential amino acids are formed. An example of this is seen in Figure 4.5.

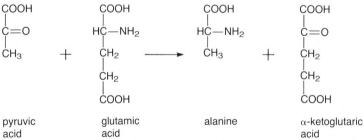

Figure 4.5 An example of the process of transamination

Urea formation

The ammonia formed from deamination is converted to urea. This is because ammonia is toxic if allowed to build up in the body, whereas urea is harmless, even in high concentrations. Urea is formed only in the liver. The reaction is basically that of uniting carbon dioxide and ammonia with the elimination of water to form urea, a relatively simple compound. However, this reaction takes place in the body not directly, but through a series of reactions called the **urea cycle**. In this cycle, the amino acid ornithine plays a major role, as shown in Figure 4.6. The amount of urea excreted will increase if the protein intake in the diet increases (see Chapter 6).

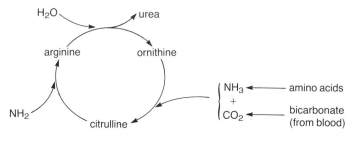

Figure 4.6 The urea cycle

4.4 Fat metabolism

Fat and catabolic products derived from fats (lipids) are used for a number of purposes in the body. All cell membranes contain the lipid cholesterol and phospholipids, such as lecithin, in addition to unsaturated fatty acids.

After absorption, fat may be oxidised by the body tissues to give energy, carbon dioxide and water, or stored in the fat deposits. Unlike carbohydrate, where there is only a limited amount of glycogen stored (0.5 kg), amounts of fat stored can be very large and variable (see Chapter 6). Carbohydrates can also be converted into fat and stored in the same way as dietary fat. On the other hand, fatty acids are not a major source of blood glucose, but rather a source of **ketone bodies** which can be used as an energy source by most tissues if glucose supply is deficient.

In starvation, or on 'reducing' diets, the fat stores will need to be called upon for a source of energy. Fat is then mobilised from the adipose tissue and passed to the liver. The neutral fat is then broken down to glycerol and fatty acids. The glycerol takes part in the glycolytic pathway and can under some circumstances be made into glucose (this is also called **gluconeogenesis** as is the formation of glucose from amino acids, see above).

The fatty acids are broken down to acetyl CoA, which can then either enter the TCA cycle (see Figure 4.4), or form **acetoacetate** in the liver by the reaction of two molecules. From acetoacetate, ketone bodies may form and are passed into the bloodstream and then to the body tissues. Most tissues, except the brain, can completely break down the ketone bodies to carbon dioxide and water in the citric acid cycle followed by the respiratory chain, with the generation of ATP as for carbohydrate metabolism. Therefore, under conditions of starvation, ketone bodies can be used as an energy source in place of glucose, although it is important to maintain the blood glucose level constant for the sake of the brain.

If tissues cannot catabolise the ketone bodies reaching them, then **ketosis** results. This is the build-up of ketone bodies in the body, which can occur in uncontrolled diabetes (see Chapter 5).

5 Major organs of metabolism: normal and abnormal functions

Body organs work in coordination with one another to maintain a constant internal environment for the existence of body cells. In this way levels of nutrients reaching the cells, and of waste products surrounding them, are maintained within quite narrow limits. The manner in which the constancy of these components is maintained is called **homeostasis**. There are many homeostatic control mechanisms in the body, but a good example is the maintenance of blood glucose level described in Section 5.6.

5.1 Endocrine glands

The endocrine glands are **ductless glands** which release **hormones** directly into the bloodstream. The function of a hormone is to have an effect on a **target organ** or cells, for example regulating growth or metabolism or the activity of enzymes. Hormones can be protein in nature, such as insulin, or steroids, such as the sex hormones.

5.2 Thyroid

The thyroid gland is the largest endocrine gland in the body, weighing between 15 and 125 g and divided into two lobes situated near to the trachea (windpipe) in the neck. The function of the gland is to produce and store the **thyroid hormones** which regulate metabolic rate. The thyroid hormones contain iodine which must be supplied in the diet. If there is too little iodine in the diet then **goitre** results (see Section 8.6).

The main effects of thyroid hormones include their metabolic effects, heat production and regulation of metabolism, and developmental effects, such as regulation of growth rate and control of protein synthesis. Protein synthesis for growth requires both thyroid hormones and **growth hormone** (see Section 5.3). In addition, thyroid hormones are also essential for normal activity of the central nervous system. They are secreted into the bloodstream and pass to all cells of the body where they gain rapid access and stimulate the DNA to increase protein synthesis.

Thyrotoxicosis (or hyperthyroidism) can occur as a result of a disease condition, or if high levels of iodine are given to people who have become accustomed to low intakes. This latter condition is largely confined to those over 40 years of age in developing countries. The condition is characterised by irritability, fatigue, anxiety, emotional instability and even tremor of the hands. It is caused by the overproduction of thyroid hormone, which leads to increased heat production by the body, associated with increased oxygen consumption. Heartbeat is accordingly increased, and there is excessive protein catabolism, bodily wasting and increased nitrogen excretion in the urine (see Section 6.4 for a discussion of nitrogen balance). The blood glucose level tends to increase, as thyroid hormones reduce the rate of insulin secretion (see below).

Abnormal conditions of the body, as found in disease, can often markedly change the body's requirement for nutrients and this is well illustrated in thyrotoxicosis. As metabolism is stimulated by overproduction of thyroid hormones, there is an increased demand for vitamins which are needed as coenzymes in some of the metabolic pathways, such as thiamin, riboflavin, B_{12} and vitamin C. In addition, there may be excessive calcium loss in urine, faeces and sweat, which may lead to **osteoporosis** (loss of calcium from the bones making them liable to break, a condition associated with old age) especially as there is less calcium absorbed from gut.

Hypothyroidism (myxoedema) in many respects represents the opposite effects of thyrotoxicosis. It is a lack of thyroid hormone, sometimes due to a disease condition. It is characterised by low blood glucose level, slowing of mental activity (speech and so on) and fatigue. It can arise as a result of disease or by a lack of iodine in the diet. The latter aspect is dealt with in Section 8.6.

5.3 The pituitary gland and the hypothalamus

The pituitary gland is the most important endocrine gland in the body. It weighs only 0.5 g and is situated in a bony cavity at the base of the skull. It lies in close proximity to the hypothalamus. The main secretions of the pituitary are **growth hormone**, adenocorticotrophic hormone (**ACTH**), **thyrotrophin** and a number of hormones which regulate reproduction and lactation. These hormones are responsible for controlling the growth, development, integrity and activity of the adrenal cortex, the thyroid gland, the ovary, breast and testis. In addition, they control the growth of bones, muscles and gut and influence the metabolism of carbohydrate, fat and protein. The hypothalamus works in a regulatory fashion. It receives messages (chemical and nervous) from many parts of the body and, as a consequence, produces **regulatory hormones**, some of which stimulate secretions of the pituitary, while others inhibit its secretion.

Thyrotrophin from the pituitary enhances uptake of iodine by the thyroid gland and increases the release of thyroid hormones to the blood. The release of thyrotrophin is regulated by negative feedback inhibition by the thyroid hormones: the lower the level of thyroid hormones in the blood, the more thyrotrophin is released.

ACTH is produced by the pituitary gland and stimulates the **adrenal cortex** to produce **cortisol** and **corticosterone** which, in turn, raise blood glucose level (see Section 5.6). The two adrenal glands are situated on the thoracic side of each kidney. Their structure is that they have an outer cortex and an inner medulla, and these two parts act as distinct organs.

The adrenal medulla is not essential for life, but can prepare the body for action following stress (the 'fight or flight' reaction in an emergency). It produces **adrenalin**, which can raise both blood pressure and blood glucose level.

Growth hormone increases the retention of nitrogen and influences growth through its control of protein metabolism. It increases the total oxygen consumption while inhibiting carbohydrate oxidation and increasing fat metabolism.

Gigantism and **acromegaly** are produced by the long-term oversecretion of the pituitary. The two disorders occur depending on the onset of the oversecretion. If it occurs before puberty, then the former case results, with people growing to 2.1–2.6 m. If oversecretion begins after puberty then acromegaly results with bone thickening and deformity and head, hands and feet and internal organs may be enlarged. There may also be increased metabolic rate and disturbance of carbohydrate metabolism (high blood glucose and glucose in urine). Insufficiency in secretion of the pituitary gland leads to a reverse of the conditions described, that is **pituitary dwarfism**.

5.4 Liver

The functions of the liver are numerous and diverse. They include the **detoxication** of toxic substances entering the body, **storage** of nutrients, formation of **blood coagulation factors**, the production of the blood protein **albumin**, the destruction of **red blood corpuscles** (as part of their natural cycle), the formation of **bile**, and numerous biochemical reactions involved in carbohydrate, fat and protein metabolism (some of these have been dealt with in Chapter 4). The importance of the liver is highlighted in many physiological textbooks, in which experiments to remove the liver of animals are described. These show a very rapid decline in blood sugar level, which can lead to coma within hours. Blood urea falls and there is a rise in blood amino acids and ammonia levels. Death, even if extra sugar is given, will occur within 18–24 hours due to the toxic effects of ammonia.

The liver is composed of **hepatic cells** which are organised into lobules. The

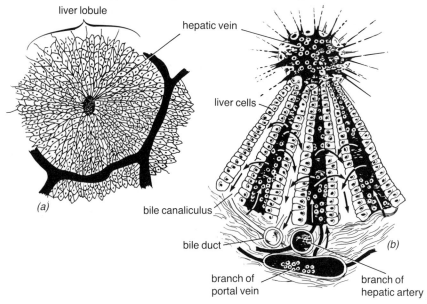

Figure 5.1 (*a*) Structure of a lobule of the liver, (*b*) detail

generalised structure of the liver can be seen in Figure 5.1. Each lobule is supplied with blood from two sources which is mixed together in the capillaries. The two sources are from the intestine, via the hepatic portal vein, and from the arterial supply to the liver from the heart, that is the hepatic artery. The combined blood then travels back to the heart in the hepatic vein.

As well as the blood supply, the lobules are supplied with **canaliculi** (bile capillaries) which form a duct system for taking away the **bile** which is produced by the hepatic cells. The bile duct system, unlike the blood supply, is a one-way system only, carrying the bile away from the cells. This system leads to the **gall bladder**, where bile can be stored and concentrated so that it can be effectively released during the digestion of a meal to aid fat absorption (see Section 3.6).

The **glucostat function** of the liver is described in Section 5.6. A certain amount of protein is stored in the liver, which can be used during fasting rather than using muscle tissue (see Section 6.4). Vitamin A and vitamin B_{12} are also stored in the liver. In the case of the former, levels in some animals can get extremely high and have been known to lead to vitamin A toxicity on ingestion (Section 6.7).

The liver protects the body from toxic substances in several ways, depending on the chemical nature of the toxin. Detoxication can occur through conjugation. In this process the toxin is combined with another molecule, such as glycine or acetic acid, and rendered harmless. Alternatively, complete destruction can take place, where the toxin is destroyed by complete oxidation. This occurs with nicotine, strychnine and barbiturate drugs.

5.5 Pancreas

As well as its function in producing pancreatic juice containing enzymes for the digestion of food (called the **exocrine** function), the pancreas also has an **endocrine** function. The cells of the pancreas which give rise to the endocrine function are present in discrete parts of the pancreatic tissue, known as the **islets of Langerhans** (Figure 5.2). These cells differ from the exocrine cells of the pancreas, called acinar cells, which are clearly situated around small ducts which will join eventually to form the pancreatic duct leading into the gut. The islets of Langerhans contain two types of cells, which can be differentiated by staining the cells with appropriate dyes. These are the **alpha** and the **beta cells**. The beta cells are responsible for the secretion of **insulin**, which is necessary for the regulation of blood glucose level (see below). The alpha cells secrete **glucagon**, which is also involved in blood glucose homeostasis.

5.6 Blood glucose regulation

The normal fasting level of blood glucose is about $80–100$ mg 100 cm^{-3} blood ($5.5–6$ mmol dm^{-3}). After eating a meal containing carbohydrate, the level rises temporarily to $120–140$ mg 100 cm^{-3}. After fasting for 24 h or more, the blood glucose level is maintained at $60–70$ mg 100 cm^{-3}. Even with a very low intake of carbohydrate, the blood glucose level does not normally fall below 60 mg 100 cm^{-3}. A low level of glucose in the blood is called **hypoglycaemia** and is as harmful to the brain as is lack of oxygen. A high level of glucose in the

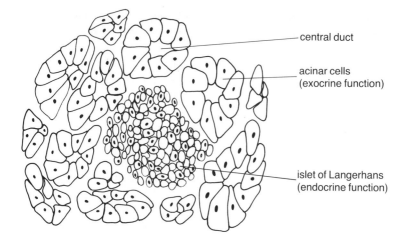

central duct

acinar cells
(exocrine function)

islet of Langerhans
(endocrine function)

Figure 5.2 Diagram of a section of the pancreas

blood is called **hyperglycaemia** and, if greater than about $180\,\text{mg}\,100\,\text{cm}^{-3}$, can lead to glucose in the urine (called **glycosuria**).

The blood glucose level at any point in time is a consequence of the rate at which glucose is entering and leaving the bloodstream. Available carbohydrates in the diet tend to raise the level, but some of this is stored as glycogen in the liver for release as and when it is required. This action of the liver in helping to maintain a constant level of blood glucose is called the **glucostat function** of the liver.

The liver is the key organ in regulating blood glucose. When the blood glucose is high, the liver takes up glucose and stores it as glycogen. When the blood glucose is low there is a net loss of glucose from the liver to the bloodstream by the breakdown of glycogen to glucose (**glycogenolysis**). Another way the liver can increase the output of glucose into the blood is by converting amino acids, lactate from muscle, or glycerol from stored fat to glucose by **gluconeogenesis**. In addition to the liver controlling blood glucose level, there is also control from several hormones. The factors which tend to lower or raise blood glucose level are shown in Table 5.1.

The most important hormonal factor tending to lower blood glucose is insulin. This lowers blood glucose in several ways, but the main one is by promoting the transport of glucose into the cells (without insulin, glucose cannot pass into the cells). Inside the cell the glucose can be oxidised, deposited as glycogen or converted to fat or amino acids, and insulin stimulates key enzymes in all of these pathways.

Hormonal factors also play a part in raising blood glucose level. Glycogenolysis (which tends to raise blood glucose) is promoted by two hormones: adrenalin and glucagon. Adrenalin is released by the adrenal glands in response to stress, and glucagon is secreted by the alpha cells of the pancreatic islets of Langerhans. Both of these promote glycogen breakdown by activating the appropriate enzyme. In long-term starvation, gluconeogenesis is a more important source of glucose than glycogenolysis, as the supplies of glycogen in

Table 5.1 Factors which tend to lower or raise blood glucose level

Lower	Raise
Fasting	Eating
Glycosuria (in diabetics)	Glycogenolysis *action of*:
Insulin	adrenalin
	glucagon
Physical activity	
	Gluconeogenesis
	Insulin antagonists:
	growth hormone cortisol

the liver rapidly reduce. Under conditions of starvation, it is normal for some gluconeogenesis to be provided by the breakdown of fat reserves (glycerol). This releases fatty acids which, together with their breakdown products the ketone bodies, can be utilised by most tissues for energy. Glucagon, adrenalin and cortisol promote gluconeogenesis in different ways as follows.

Glucagon – stimulates enzymic pathways from amino acids to glucose.
Adrenalin – inhibits the secretion of insulin.
Cortisol – stimulates synthesis of enzymes responsible for gluconeogenesis.

Thus, these hormones act in concert to increase blood glucose level, partly by increasing glucose inflow into the blood and partly by reducing its outflow into the tissues.

The condition of **diabetes** is characterised by a high blood sugar level and an excretion of glucose in the urine in the untreated state. It is often caused by an insufficient secretion of insulin by the pancreas. At the same time, there may be an excess of glucagon which raises the blood glucose level. Although the blood glucose level is high, the glucose cannot enter the cells because of the lack of insulin and, therefore, the untreated diabetic shifts from carbohydrate metabolism to fat metabolism for energy, which results in the formation of ketone bodies. In severe cases this can cause acidosis with the formation of acetone which can be detected on the breath.

The most commonly used test for diabetes is the **glucose tolerance test**. To carry out this test, glucose is given orally after fasting, at a dose of 1 g per kilogram body weight. In the normal person the blood glucose level rises from about $90\,mg\,100\,cm^{-3}$ to $140\,mg\,100\,cm^{-3}$ and then falls to normal within about 3 h. This fall comes from the release of insulin, following the elevation of blood glucose level. In the diabetic, the rise is much greater and will often greatly exceed $180\,mg\,100\,cm^{-3}$, at which level glucose normally appears in the urine. This level is called the **renal threshold**. The fall is much slower in the diabetic and in some cases even the resting value may be as high as $300\,mg\,100\,cm^{-3}$. Proper treatment of the diabetic depends on the type of diabetes. The more severe **juvenile** type or insulin-dependent type needs the administration of the right amount of insulin by injection. The **mature-onset** or non-insulin-dependent type may be treated by adjustments to diet. These days, diabetics are advised to eat a diet high in starch and dietary fibre, to avoid the effects of the high-fat diet previously prescribed which increases the risk of coronary heart disease. In fact, the diet for diabetics is very similar to the dietary guidelines being advocated for the general population (see Table 1.5).

6 Nutrient balance

6.1 Introduction

Recently, with the publication of two reports (NACNE, 1983 and COMA, 1984) on health and diet, attention of the media has moved away from considerations of adequate nutrition to a preoccupation with the amounts of fats, sugar and dietary fibre in the diet. While it may be true that in the United Kingdom more illness could be caused by overeating than by a deficiency in nutrients, nevertheless, healthy nutrition starts with an understanding of the body's requirements for nutrients. Despite overeating, it is still possible to be deficient in some nutrients, as the intake of food in humans is determined by the energy content of the diet and not by its content of other nutrients (sometimes referred to as **nutrient density**). In this chapter maintenance of body balance of major and minor nutrient components of the diet will be examined.

6.2 Energy

Energy of the diet is used to perform mechanical work, to maintain the tissues of the body and for growth. Despite protestations from those who cannot lose weight that they do not eat but merely live on thin air, biochemical processes in humans follow the laws of thermodynamics applicable to other chemical reactions. Therefore, humans can neither create nor destroy energy. The human body is a relatively poor converter of energy, only about 25% of energy from food gets converted to mechanical energy. All energy forms produced in the body are ultimately converted to heat (with the exception of the potential energy stored in the pyramids of course!). Advantage is taken of this fact in measuring energy expenditure of the body using a **human calorimeter** (see Section 6.3).

From Chapter 4 it can be seen that the breakdown products of proteins, as well as fats, can be used as an energy source. The use of proteins as an energy source is not often appreciated especially by those who consider proteins to be 'slimming'. Most well-fed humans eat more protein than they need for body-building purposes or tissue repair and this extra protein is used as an energy source on a daily basis. Problems arise, however, if the body does not take in enough energy from carbohydrate and fat. In these circumstances, protein in the diet will be channelled directly for use as an energy source, by-passing its normal function for growth and maintenance, effectively depleting the body of protein. This sad state of affairs occurs in protein energy malnutrition (Section 8.5).

Units of energy

Attempts have been made in recent years to bring nutritionists into line with the use of SI units. This means replacing the **Calorie**, with which many people are familiar, by the **joule** as the unit of energy. Dietitians are reluctant to move over to the new units when dealing with patients, as the Calorie seems to be one of the few concepts in nutrition which a large proportion of the public understand. In the eyes of the public, there is too much contradictory information on diet already. Therefore, at the moment, nutritionists tend to publish original work in joules, but to think and communicate in popular terms in Calories (or kilocalorie).

Confusion also arises with the term 'calorie'. The calorie (with a small c) is a unit used in physics and is the amount of heat required to raise the temperature of 1 cm³ of water from 15 to 16 °C. This unit is too small to use in human nutrition, and so the kilocalorie (kcal) is used instead, which is the amount of heat required to raise the temperature of 1 dm³ of water from 15 to 16 °C. Therefore 1 kcal = 1000 calories.

The trouble comes, because in everyday parlance the kilocalorie is called the 'Calorie' (with a capital C), but magazine writers and others are not aware of this nuance and therefore write 'calorie' when they mean 'Calorie'. To avoid confusion it is better to use the term kilocalorie.

A joule is the energy expended when 1 kilogram is moved 1 metre by a force of 1 Newton. 1 megajoule (Mj) = 1000 kilojoules and 1 kilojoule (kj) = 1000 joules. In human nutrition it is the kilojoule and megajoule which are used, as the joule itself is, again, too small to use.

6.3 Energy expenditure and energy intake

Unsurpassed experimentation in this area was carried out by Atwater and his colleagues at the turn of the century, using a human calorimeter (see Plate 1). Atwater's experiments were perhaps the most detailed and careful of any experiments recorded in the history of nutritional science. Using the human calorimeter, Atwater confirmed that (*a*) energy expenditure was directly related to oxygen consumption, and (*b*) total energy expenditure was equivalent to the energy content of foods if digestibility and lowered utilisation of protein by the body were taken into account.

Atwater and his colleagues constructed a box big enough to hold a person. Pipes carrying water were fed into and out of this chamber and the temperature of the water could be noted on entry and exit. From these measurements the amount of heat given out by a person could be calculated. At the same time the amount and composition of the air going into the chamber could be measured.

Atwater's conclusion, that energy expenditure was related to oxygen consumption, lead to **indirect calorimetry**, a much more convenient way of determining human energy expenditure by merely measuring the volumes of oxygen used during any activity. Several pieces of apparatus are available for this, but perhaps the most flexible is the Max Planck respirometer which is like a gas meter. If connected to the mouth by tubing via a one-way valve, the volume of the expired air can be recorded. A sample of the air for oxygen analysis can be taken at an appropriate time into a bladder on the equipment.

The **energy intake** of a person can be determined by measuring the energy content of the diet. The **energy content** of a food can be determined by burning it

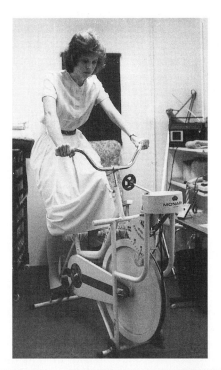

Plate 1 A human calorimeter; (*a*) inside view, (*b*) sophisticated instrumentation required for measuring heat production and oxygen consumption

in an atmosphere of oxygen in a **bomb calorimeter** and measuring the heat produced. If 1 g samples of fat, protein, carbohydrate or alcohol are burnt in a bomb calorimeter, then they will give out different amounts of heat (Table 6.1).

Table 6.1 The heat of combustion and the available energy in protein, fat, carbohydrate and alcohol

	Heat of combustion/ kcal g⁻¹	Loss in urine/ kcal g⁻¹	Availability/ %	Atwater factors/ kcal g⁻¹
Protein,				
egg	5.58	1.25	92	4
Fat,				
butter	9.12	—	95	9
Carbohydrate,				
starch	4.12	—	99	4
glucose	3.69	—	99	4
Ethyl alcohol	7.10	trace	100	7

Fat contains a lot of energy, as does alcohol (a point often overlooked by slimmers). Table 6.1 also indicates the amount of heat produced in the body by 1 g of these dietary components. These values are called the **Atwater factors** as they were determined using a human calorimeter by Atwater on volunteers eating diets composed of single substances. It will be seen that the values for fat, carbohydrate and alcohol are quite similar to that obtained in a bomb calorimeter, the small differences being due to small losses of these substances in the faeces thus lowering the availability. (If all fat and carbohydrate were absorbed in the gut then availability would be 100%.) However, the values for protein are quite different. This is a reflection not only of some loss of protein in the faeces, but also the fact that protein in the body breaks down only so far, to **urea**, which is excreted in the urine. If you placed pure urea in a bomb calorimeter then this would be capable of being burnt to produce heat and gaseous nitrogen oxides as end-products.

Atwater factors can be used to calculate the energy value of a food from its composition. An example, the composition of baked beans in tomato sauce, is analysed in Table 6.2.

Table 6.2 Calculation of the energy content of baked beans using Atwater factors

Component	%	Atwater factor		Energy value kcal
Water	74			
Carbohydrate	10	× 4	=	40
Protein	5	× 4	=	20
Fat	0.5	× 9	=	4.5
Totals	89.5*			64.5

*This value does not add up to 100 because dietary fibre and minerals are not included.

6.4 Protein

Protein is the most abundant component of the body after water. Half of the dry weight of the body is protein. It is required for the formation of new tissues during growth. Tissues are also laid down during pregnancy for the foetus and placenta and when forming new muscular tissue during physical training.

As well as body building, protein is required for tissue replacement or repair. Although some cells, such as those of the nervous system, are never replaced in a lifetime, the proteins within them are still turning over. This is called the **dynamic state of body proteins** and involves **anabolism** (synthesis of proteins) and **catabolism** (breakdown of proteins). Therefore, the requirement for protein is divided into two: the **growth** and the **maintenance** requirement.

Proteins are composed of amino acids and some of these must be supplied as nutrients in the body, that is the **essential amino acids**. Others can be produced in the body from **keto acids** by a biochemical process called **transamination** (see Section 4.3). The essential amino acids normally have a characteristic feature which would make them difficult to manufacture within the body. Some of these can be seen in Figure 2.7, contrasted with simple amino acids such as glycine which can be synthesised within the body.

Human protein requirements

The intake and output of the protein in the body can be measured by the **nitrogen** intake and output, or **balance**. On average, proteins contain 16% nitrogen (the usual factor for conversion of nitrogen to protein is to multiply the nitrogen value by 6.25, which is 100 divided by 16). The reason for measuring nitrogen is that it is fairly easy to measure, and is not present in fats and carbohydrates which are the other main components of the body and the diet. However, the diet does contain some non-protein nitrogen which can be quite considerable in some foods. Allowance can be made for this, by using modified factors for converting nitrogen to protein for these foods, although it is more common for nutritionists to ignore them when studying nitrogen balance.

Human protein requirements can be measured by nitrogen balance studies (Chapter 1) whereby a person is put onto a diet low in protein, when the nitrogen balance will be negative. In other words more nitrogen will be lost from the body per day than is being taken in. The intake of protein can then be raised until a minimum amount of nitrogen is present in the diet to achieve nitrogen balance (intake = output). This amount of protein per day can be regarded as the **physiological requirement** of a protein.

For protein, there is considerable difference between RDA (recommended daily amount) values from different countries (see Appendix for UK values). However, they are all based on the same human data, published internationally. Differences arise because of the different interpretations of what an RDA is, and differences in philosophy of the various committees who sit to deliberate on these matters. Thus, in the UK the protein RDA is set at an arbitrary figure of 10% of the food energy by the DHSS (1979) who take the view that a diet providing less than 10% of total food energy as protein is likely to be unpalatable to most people in the UK and may be deficient in other nutrients which are associated with protein.

On the other hand, the international RDAs (FAO, 1973) are based on the physiological requirements, with the addition of a 'safe level' which is an arbitrary amount of 30% extra. FAO in 1973 took the view that, as there is no known deleterious effect for healthy people of too much protein intake, it is better to err on the high side rather than risk people being deficient. Nevertheless this method of calculation gives much lower RDAs than the DHSS approach as can be seen in Table 6.3.

Table 6.3 Comparison of RDAs for protein for various human groups (g person^{-1} day^{-1})

	DHSS (1979)	FAO (1973)
Children 4–6	43	20
Male adolescents 13–15	66	37
Female adolescents 13–15	53	31
Men	72	37
Women	54	29

6.5 Water

The water content of an adult man is about 40 dm^3, the loss of about 2 dm^3 of which would cause discomfort, 4 dm^3 would lead to disability and 8 dm^3 would be fatal. In the body, water functions, among other things, as a solvent, takes part in body building (for example, is required if protein is laid down in body tissues), acts as a catalyst in some body reactions and as a lubricant.

In a temperate climate, water losses from the body amount to about 800 cm^3 from skins and lungs per day. On the other hand, minimum losses from kidneys are 300–500 cm^3 per day. As the body produces about 200–300 cm^3 per day from oxidation of energy sources, the dietary requirement for water is at least 100 cm^3 per day in a temperate climate. As losses from skin and lungs can increase by a factor of four in very hot atmospheres or with heavy physical work, more than 1 dm^3 per day would be required under such conditions.

6.6 Minerals

The amounts of most minerals in the diet of most well-nourished people are adequate to maintain balance. Therefore the only minerals in the UK for which RDAs are given (DHSS, 1979) are calcium and iron and, as space is short, comments on mineral balance will be confined to these. As with most minerals, calcium and iron balance of the body is not only affected by the amounts present in the diet, but by a whole range of factors which influence their absorption. These factors have already been dealt with in Sections 3.9 and 3.10.

Iron

Adults have 3–4 g of iron in their bodies, of which two-thirds is present as **haemoglobin** in the red cells of the blood. Haemoglobin carries oxygen from the lungs around the body. After absorption of iron from the food by the mucosa of the small intestine, the iron is bound to **globulin, transferrin** or **apoferritin** within the mucosa. The former two are part of the body's transport system for iron. If all the globulin and transferrin have been used up by binding with iron, then further iron binds to apoferritin to form **ferritin** within the mucosal cells and is excreted when the cell is shed into the lumen of the gut at the end of its three-day life. In this way the body avoids absorbing too much iron. Apart from this, other losses of iron by the body include that in the urine, although this is normally quite small. In addition, women may lose iron during menstruation, at birth of their infants (either to contribute to the child's iron store or as haemorrhage at birth) and during lactation. Children may have extra demands during growth. The RDA of iron for women is higher than for men (Appendix), and with most diets containing an average of only about 6 mg 1000 kcal^{-1}, the diet may not permit some women to maintain an adequate iron balance.

Lack of iron causes **iron deficiency anaemia**, which is common in women and children worldwide. It is characterised by a low level of haemoglobin in the blood, which is not in itself life threatening, although it can lead to considerable debility. In Third World countries, iron deficiency anaemia is often caused by a malarial infection. People in such areas may become infected with malaria from *Anopheles* mosquito bites. The malarial parasite causes anaemia because, during malarial fever, the parasite causes breakdown of the red blood cells, the haemoglobin is released and broken down and iron is lost in the urine.

Calcium

Calcium is an essential nutrient which must be provided regularly in our diet. In all, the body of an adult contains about 1200 g of calcium, with most of that in the skeleton where it is deposited together with phosphorus as **hydroxyapatite** crystals. Calcium is also necessary for normal blood clotting, transmission of nerve impulses, muscle contraction, hormone secretion and enzyme activation. Indeed, it is present in small amounts in all fluids of the body and this is maintained in equilibrium with that of the bones under normal conditions (see Figure 6.1).

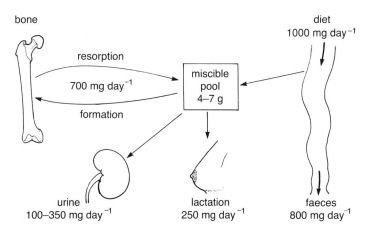

Figure 6.1 Calcium exchanges in the body

Factors causing poor absorption of calcium have already been discussed in Section 3.9. The excretion of calcium is relatively straightforward compared with other minerals as it is simply excreted in the urine.

Most essential nutrients are found in a variety of foods, but in the case of calcium we rely heavily on one dietary source: milk and its derivatives cheese and yogurt. One pint of milk contains 0.6 g of calcium. Overall, milk and other dairy foods provide 56% of the calcium intake in the UK. A further 25% comes from cereal foods and 7% from vegetables.

Adults are normally in calcium balance, that is the amount of calcium in the faeces and urine equals that in the diet. However, children who are growing are normally in positive balance, when intake exceeds output and calcium is being accumulated within the body. Negative calcium balance, where output exceeds intake, can lead to lack of calcium in the bones and the bending of leg bones under the weight of the body. This condition is called **rickets** if it occurs in childhood and **osteomalacia** if it occurs in adulthood. These forms of calcium deficiency are normally associated with low vitamin D status of the body. Vitamin D aids calcium absorption and is essential for proper calcium balance. **Osteoporosis** is loss of calcium in the bones of the elderly and is more complex in its development (see Section 7.5).

6.7 Vitamins

One way of determining the adequacy of vitamin intake into the body is to measure the level of the vitamin in the blood. However, although this measure will indicate whether a person is deficient, it will not give any indication of the amount present in the body stores. In healthy well-nourished people, although stores of vitamin A and B_{12} are held in the liver, there are no substantial stores of other vitamins in the body. In contrast to water-soluble vitamins, excess intakes of all the fat-soluble vitamins have the potential to be accumulated within the body to very high levels (and are therefore potentially toxic), as these vitamins are not excreted in the water-based environment of the urine as are the water-soluble ones.

It is not possible in this short review to give a meaningful description of the factors affecting the balance of all vitamins within the body and therefore two have been selected for special mention. Both are important from the point of view that they can be low or marginal in the diets of some people. Vitamin C is a water-soluble vitamin, particularly susceptible to loss during processing (see Chapter 9) as it is the first nutrient which is destroyed during food processing. Vitamin A is a fat-soluble vitamin in short supply in the diets of many children in the poorer areas of the world, where deficiency can lead to blindness.

Vitamin C

Vitamin C is commonly low or marginal in diets all over the world. In many Third World countries, lack of the vitamin is linked with iron deficiency anaemia, as vitamin C is one of the factors aiding iron absorption. In many respects, vitamin C typifies the behaviour of water-soluble vitamins in being easily absorbed and excreted from the body, either as the active vitamin or its breakdown products. As human beings, we have a special requirement for vitamin C, as, unlike most animals, we cannot make our own as we do not produce the enzyme **gulonolactone oxidase** which is necessary in the synthetic pathway. This inability to produce vitamin C is a problem for only a few animals, including primates, guinea pigs, carp and trout.

In the body, vitamin C is involved in the synthesis of **collagen**, (an important structural connective tissue which constitutes about 25–33% of the total protein of the body), tissue proteins, **lipoproteins** (proteins with a lipid part, which are used in the transport of lipid in the aqueous medium of the blood) and blood plasma. In the synthesis of collagen, vitamin C is necessary for the formation of **hydroxyproline** from the amino acid proline; hydroxyproline constitutes about 15% of the protein which goes to make up collagen. No other protein has such a high proportion of hydroxyproline.

Deficiency of vitamin C leads to anaemia (low blood haemoglobin level) and **scurvy**, which is characterised by bleeding gums, haemorrhages under the skin and slow wound healing. Although the RDA for vitamin C in the UK is only 30 mg per day for adults (see Appendix) which is adequate to prevent deficiency symptoms, some people have advocated higher doses for extra health benefits. The **megadose hypothesis** of vitamin C applies to doses greater than $500\,\text{mg day}^{-1}$ up to the level which the gut of the individual will tolerate without showing signs of diarrhoea, that is doses in excess of $7\,\text{g day}^{-1}$. These large doses are said to be protective against a number of diseases, including the common cold, cancer, and arterial disease (and therefore heart disease). Vitamin C is said to promote the action of the immune response and thus improve the body's natural defence mechanisms. However, results of experiments with viral infections and cancer have not yet yielded unequivocal evidence in favour of the protective function of vitamin C for these diseases. Nevertheless, there is sufficient support in the USA for the concept of vitamin C being useful in the prevention of cancer to lead to a strong lobby in favour of increasing the RDA for vitamin C for this reason.

As far as arterial disease is concerned, there is evidence that megadoses of vitamin C promote formation of bile acids from cholesterol and therefore lower blood cholesterol level. Recent studies using guinea pigs have shown that animals that are vitamin C deficient have higher incidence of **atherosclerosis** (furring up of the arteries which, in humans, is a condition preceding coronary heart disease, see Section 8.2). It has also been shown in humans with high levels of blood cholesterol (which means high risk of coronary heart disease) that high doses of vitamin C can reduce blood cholesterol level, especially if the patient normally eats a diet with a low content of vitamin C.

Vitamin A

All vitamin A comes indirectly from plant carotenoids. There are about ten **carotenes** which can be considered as **provitamin A** as they can be converted to vitamin A in the body. This mainly occurs in the intestinal mucosa, but also in the liver and kidneys. Conversion is always incomplete, but beta carotene is more effectively converted to vitamin A than the other carotenes.

Carotenes are associated pigments of chlorophyll, which help to trap energy from sunlight. Therefore, the greener the leaf, the more chlorophyll it will contain and, therefore, the higher the content of carotenes. Carotenes are especially high in spinach and members of the *Brassica* family. The pre-formed vitamin A is only found in foods of animal origin. The consequences of diets low in vitamin A or carotenes are discussed in Section 8.7.

Very large quantities of fat-soluble vitamins are required to cause **toxicity**, but despite this there are cases of human toxicity due to vitamin A or the carotenoids reported in the literature. In the marine environment, the main sources of the vitamin are minute plant life, such as diatoms, which contain carotenes. Animals that eat the plant life convert the carotenes to vitamin A, and the vitamin A becomes concentrated in the food chain. Consequently, animals eating fish high in the food chain eat large quantities of vitamin A. There are a number of reports of the high vitamin A in polar bear liver poisoning arctic explorers.

6.8 Essential fatty acids

The necessity for fatty acids in the diet was first noticed by two workers from California, George and Mildred Burr, in the USA in the 1930s. They fed rats fat-free diets and found that they developed a whole range of symptoms, including skin lesions, decrease in body weight, and heart and kidney enlargement. As their discoveries shortly followed the discovery of the main vitamins, they called the component of fat which would alleviate this condition 'vitamin F'. Nowadays we call these factors (as there is more than one) **essential fatty acids (EFA)**.

Although the fatty acid identified by Burr and Burr was **linoleic acid**, we now know that there are a number of fatty acids that show EFA activity to varying degrees, but they all have some features in common. They are all polyunsaturated and the unsaturated bonds have to be in specific places in the molecule. EFA have double bonds in the n-6 and n-9 positions, as shown in Figure 2.14.

Deficiency of fatty acids in the diet is never seen in humans, because requirements are estimated at between 2–10 g linoleic acid per day, which is present in most diets, especially if that diet contains a high proportion of foods of plant origin. Although linoleic acid is by far the most common polyunsaturated fatty acid in our diet, there are others present with EFA activity. Table 6.4 shows the fatty acid composition of some fats used as food.

Table 6.4 Common fatty acids in some foods (% of total fat)

Fatty acid	Number of carbons	Number of unsaturated bonds	Beef muscle	Beef fat	Butter	Corn oil	Olive oil	Green leaves
Palmitic	16	0	13	25	30	13	9	13
Stearic	18	0	16	29	11	3	3	0
Oleic	18	1	21	34	19	31	77	7
Linoleic*	18	2	20	2	2	53	11	16
Linolenic*	18	3	2	1	0	0	0	56
Arachidonic*	20	4	19	0	0	0	0	0

* polyunsaturated or essential fatty acid.

Essential fatty acids are used by the body for the formation of cell membranes. Other functions include the formation of **prostaglandins** in the body. Firstly, the liver converts linoleic acid to another fatty acid called **arachidonic acid**. This is then transported from the liver to the tissues of the body where prostaglandins are synthesised. The various prostaglandins have a wide range of activity at very small concentrations. For example, they influence the function of the central nervous system, contraction of involuntary muscle such as that of the intestine, and influence blood pressure. Because they are so powerful, it is important that they are made at the site of action and destroyed very quickly. Hence their half-life is very short.

Since the discovery of prostaglandins, other products from the EFA have been discovered, namely the **throboxanes** in the blood platelets which have anti-platelet aggregating activity. Other effects depend on the particular thromboxane, and include contraction or relaxation of smooth muscle, promotion of platelet aggregation, relaxation or constriction of the muscles of the blood vessels causing a raised or lowered blood pressure.

The amounts of EFA necessary to cover requirements are usually considered to be small. It is a matter of controversy whether larger intakes are beneficial. However, there is no doubt that, as well as the effects of EFA described, the addition of different fats to the human diet has been demonstrated to change the levels of cholesterol and triacylglycerols in the blood. Table 6.5 shows the effects of feeding various pure fatty acids in trials using humans. As raised levels of these substances have been associated with heart disease, recommendations for lowering the amount of saturated fats in the diet and increasing the P/S ratio (polyunsaturated:saturated fatty acid ratio) of the diet have been made (Chapter 1).

Table 6.5 Effects of pure fatty acids on blood cholesterol and triacylglycerols in human trials

Fatty acid	Number of carbons	Number of unsaturated bonds	Plasma Cholesterol	Plasma Triacylglycerols
Medium chain	8–10	0	0	↑
Lauric	12	0	↑	0
Myristic	14	0	↑↑	0
Palmitic	16	0	↑↑	0
Stearic	18	0	↑	↑
Oleic	18	1	0	0
Linoleic	18	2	↓	↓

↑, increase; ↓, decrease

6.9 Dietary fibre

Strictly speaking, the subject of dietary fibre should not appear in a chapter entitled 'nutrient balance', as little nutritional benefit is derived from fibre itself. However, the presence of fibre in the diet has considerable influence on the balance of other nutrients present, so it is convenient to consider it here with the discussion of the other nutrients still fresh in mind.

The present upsurge of interest in the benefits of dietary fibre in human nutrition owe much to the work of two medical doctors, Denis Burkitt and his colleague Trowell. They both worked in Africa making careful observation of the differences in diseases patterns suffered by Africans compared with westerners. The study of disease patterns within and between communities in this way is called **epidemiology**. Epidemiology cannot prove a relationship between cause and effect, because there are so many unknown factors (variables) which cannot be taken into account in any study and therefore are not controlled. One of the problems of studying the effects of a high fibre diet on disease patterns is that if you change the fibre content of the diet, particularly by advocating eating greater quantities of less refined food rather than using isolated fibre preparations, you automatically change other aspects of the diet. This is particularly so if you are conducting the studies on free-living individuals and make use of commonly encountered foods. Thus, it may be very difficult for researchers to distinguish between the effects of dietary fibre and the effects of other changes in the diet.

Characteristics of a high fibre diet are summarised in the Table 6.6.

Table 6.6 Characteristics of diets rich in dietary fibre

Less dense (bulk) higher volume/weight
Less energy dense, lower kcal per gram
Low in fat, refined sugar (and salt)
Rich in starch
Protein predominately of vegetable origin rather than animal

Despite the problems of interpretation of epidemiological observations, they can indicate possible relationships to form the basis of an hypothesis. It is for this purpose that epidemiology has been used in the study of dietary fibre in human nutrition. Further work is always needed to confirm hypotheses based on epidemiological observations in the form of animal studies or clinical trials, and for dietary fibre many of these are currently underway. Information based on these types of studies is gradually emerging and so far tend to support the 'dietary fibre hypothesis', which can be summarised as:

dietary fibre is protective against 'diseases of affluence',

low fibre diets are a causative factor in the development of some diseases (such as diverticulitis, colon cancer).

Diverticular disease was one of the first diseases studied by Burkitt in relation to dietary fibre. The condition results from passing hard, dry stools over a long period of time. On a high fibre diet the stools are soft and bulky and readily passed through the large intestine, propelled by segmentation. On a low fibre diet, the segments which are formed are closed off completely and a high pressure develops inside them, which, over the years, tends to push the mucosa through the thick muscular wall of the gut to form a structure rather like a small balloon containing faecal material. Human beings can have many of these diverticulae without any symptoms, but larger numbers do give pain and can act as focal points for infection, leading to diverticulitis. In the past this condition has been treated with bland diets and surgery, but, increasingly, high fibre diets are used with great success and in many cases avoid the necessity for surgery.

Other diseases associated with low fibre diets are summarised in Table 6.7. Further discussion of the role of dietary fibre in blood glucose control for diabetics is given in Chapter 7.

Table 6.7 Diseases associated with low fibre diets

Constipation
Diverticular disease
Cancer of the colon
Appendicitis
Varicose veins, haemorrhoids and venous thrombosis
Atherosclerosis and coronary heart disease
Diabetes
Peptic ulcers

6.10 Dietary fibre: definitions, analysis and sources

There are two main ways of defining fibre:

physiological: the sum of lignin and the polysaccharides that are not digested by the endogenous secretions of the human digestive tract;

chemical: the sum of lignin, cellulose and hemicellulose.

Problems of definition partly account for wide variation in reported values for the dietary fibre content of foods. In older books dealing with diet, it is common to come across the term **crude fibre**, which should not be confused with dietary fibre. Crude fibre methodology was devised for understanding the nutrition of the cow and, applied to foods, grossly underestimates fibre for human nutrition. The determination of dietary fibre is by no means straightforward. There are **fractionation** methods whereby the food is broken down gradually by reagents of increasing stength and at each stage the breakdown products are determined. These methods yield a lot of detailed information of the fibre components but are complex and time-consuming.

Other methods are called **gravimetric**, in which reagents are used to remove all substances except the dietary fibre, which is then dried and weighed. Reagents such as gut enzymes or detergents can be used for this purpose, and this has given rise to the Acid Detergent Fibre (**ADF**) method, which determines cellulose and lignin, and the Neutral Detergent Fibre (**NDF**) method, which determines cellulose, lignin and hemicellulose. Recent refinements of both fractionation and gravimetric methods have included the division of fibre into 'soluble' and 'insoluble' fibre. Soluble fibre seems a contradiction of terms, as the concept of most fibres is that they are insoluble in water. In fact soluble fibres are not even fibrous, they comprise the gummy substances like pectin, which is used for its setting properties in jam-making.

Dietary fibre is of plant food origin. As it is found in the plant cell wall as pectins or hemicelluloses and cellulose, all plant sources contain some fibre, but some sources have a much higher content than others. Wholemeal cereals are particularly high, as the bran around the seed grain contains large quantities. Another outstanding source of dietary fibre is the seeds of the pea and bean family Leguminoseae, most commonly eaten as the canned baked bean in the UK. Most leafy vegetables and fruit are disappointingly low in fibre, mainly because they are so high in water. Some sources of dietary fibre are given in Table 6.8.

Table 6.8 Some sources of dietary fibre, showing amounts required to provide 30 g per day (in descending order)*

Food	Water	Dietary fibre	Amount required to give 30 g as eaten	Dietary fibre (dry weight basis)
	%	%	g	%
Tomatoes, raw	93.4	1.5	2000	22.7
Strawberries, raw	88.9	2.2	1364	19.8
Rhubarb, raw	94.2	2.6	1154	44.8
White bread	39.0	2.7	1111	4.4
Carrots, old	91.5	3.1	968	36.5
Peas, garden, canned	81.6	6.3	476	34.2
Spinach, boiled	85.1	6.3	476	42.3
Haricot beans, boiled	69.6	7.4	405	24.3
Peanuts, fresh	4.5	8.1	370	8.5
Wholemeal bread	40.0	8.5	353	14.2
Cornflakes	3.0	11.0	273	11.3
Weetabix	3.8	12.7	236	13.2
All Bran	2.3	26.7	112	27.3

* from Paul & Southgate, 1978.

Although there are few guidelines at present on the intake of dietary fibre, the NACNE Report recommended increasing intake to 30 g from 20 g per day. The increase should come mainly from the increased consumption of wholegrain cereals.

6.11 Dietary fibre: its nutritional contribution

The main components of dietary fibre are described in Section 2.1. The principal components are carbohydrates. Until recently, it was assumed that dietary fibre makes no great contribution to energy intake. However, it has now been established that hemicelluloses can be fermented in the large intestine to produce **short-chain fatty acids** which are absorbed in the large intestine and used by the body as an energy source. For some dietary fibre sources, up to 50% of the fibre may be utilised in this way. The presence of a high dietary fibre content of the diet can affect the absorption of other dietary components, especially protein, fat and minerals. The lowered availability is reflected in small changes in the Atwater factors (for a discussion of Atwater factors see Section 6.3). Thus, the Atwater factors are 3.6 instead of 4 for protein and 8.8 instead of 9 for fat, with only very small changes for carbohydrate.

The other way in which it is suggested that a high fibre diet makes a nutritional impact is the **satiety effect** on the satiety centre of the brain. With high bulk and low energy contribution it may sound like good common sense to suggest that a high fibre diet can help you lose weight. But there is very little scientific evidence to back up this hypothesis at the moment, although work is progressing. The main reason for this is difficulty of working in the area, which impinges on psychology.

A serious drawback to high fibre diets is the **lowered bioavailability of minerals**, because of the mineral binding effect of various fibre components. The minerals which are of particular concern are those which are present in the diet in low or marginally adequate amounts, such as calcium, zinc and iron. However, there is an added complication here as, for seeds such as cereals and legumes, it is very difficult to distinguish the binding of minerals to fibre from the binding of minerals to **phytate** (see Section 3.9). Nevertheless, it is now clear that dietary fibre itself binds minerals, but the extent of binding depends on the minerals concerned and the type of fibre. Various studies have given conflicting information on the relative importance of dietary fibre binding, or phytate binding, of minerals. In some studies with diets fed to humans over long time periods there is evidence of adaptation to high fibre diets and the utilisation of minerals is almost the same as without the fibre. However, other similar studies show low bioavailability of minerals on high fibre diets, with no body adaptation over time.

This is an area of a lot of active research at the moment but, until more is known, it should be assumed that high fibre diets lower mineral utilisation and that the intake of minerals of those who eat high fibre diets should be increased. We do not know the extent of body adaptation to a high fibre diet, but we can be consoled by the fact that only a few overt cases of mineral deficiency due to high fibre diets have been reported.

6.12 Physiological effects of dietary fibre

The most obvious effects of dietary fibre on the human body are to **increase stool weight** and to **decrease transit time** (time for food to pass through the body). These effects have been demonstrated over and over again. An example of these studies is shown in Figure 6.2, which shows an inverse relationship between stool weight and transit time.

As dietary fibre is a variable commodity, depending on the source, it is not surprising that different dietary fibre sources act differently in increasing stool weight and decreasing transit time. The increase in stool weight is largely due to the extra amount of water which is bound by the cellulose component in the

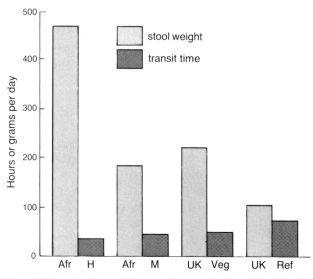

Figure 6.2 Intestinal transit times and average daily weight of stools in a group of volunteers on high- and low-fibre diets: Afr H, African villagers on an unrefined diet; Afr M, Africans on a mixed diet; UK Veg, UK vegetarians on a mixed diet; UK Ref, volunteers on a refined UK diet

large intestine. Figure 6.3 shows the effects of various fibres in increasing stool weight.

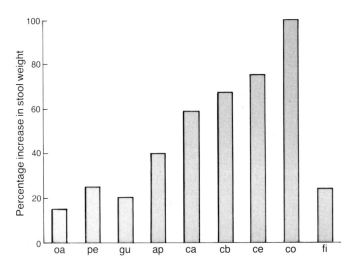

Figure 6.3 Effects of various dietary fibres on increasing stool weight: ap, apple; ca, carrot; cb, cabbage; ce, cellulose; co, coarse wheat bran; fi, fine wheat bran; gu, guar gum; oa, oat bran; pe, pectin

Besides these two principal effects of dietary fibre, other physiological effects include lowering cholesterol level and blood glucose level (the latter is particularly shown by legume fibre and provides the main basis for the better control of diabetics through high-fibre diets (see Section 7.8).

7 Dietary requirements of various human groups

The main differences in RDAs for most minor nutrients are between those for adults and those for infants/children less than 4 years old (see Appendix), otherwise they do not vary greatly from one human group to another. The reasons for this are partly because of incomplete knowledge about factors influencing requirements of vitamins and minerals. Much more is known about energy and protein requirements and therefore RDAs for these are much more detailed. Energy intake influences the intake of other nutrients which are carried along with energy in the diet. For this reason, factors influencing energy requirements are particularly considered here with those of protein for comparison.

7.1 Factors influencing energy and protein requirements

Requirements for energy and protein are influenced greatly by age and sex. These two factors are incorporated in RDA tables, but other influences such as disease (Table 7.1) are not.

Table 7.1 Main factors influencing energy and protein requirements

Energy	Protein
. . . . .Growth/age.	
. . . . .Body size and composition.	
. . . . .Pregnancy.	
. . . . .Lactation.	
. . . . .Disease.	
Basal metabolic rate	
Diet-induced thermogenesis	
Physical activity	
	Injury
	Energy intake
	Protein quality

The most important factors determining the energy requirement of an individual are **basal metabolic rate** (**BMR**), diet-induced thermogenesis (called specific dynamic action in older books) and physical activity. BMR constitutes about three-quarters of our requirement for energy and is required for the maintenance of chemical processes in the body, heart beat, respiration and maintenance of body temperature.

BMR is measured as the heat loss from the body, or indirectly as the oxygen consumed in a given time. The person must be lying down in a warm room, 12 h after eating and at complete physical and mental rest. Two of the factors affecting BMR are age and gender (because of the body compositional differences between male and female, that is 15 and 25% fat respectively for normal weight people). In late pregnancy (last three months) there is an increase of 20% in the BMR.

Factors increasing BMR are diseases (13% increase in BMR for each degree Celsius rise in feverish conditions) and over-active thyroid gland (hyperthyroidism). Although there are no racial differences in BMR, those living in very hot climates may reduce their BMR by about 10%. Other factors which reduce BMR are fasting and low body weight (between 20–25% in late stages of starvation) which may be an adaptive mechanism of the body to conserve energy.

Diet-induced thermogenesis (DIT)

After food is eaten there is an increase in heat production by the body. The increase is greater if pure protein is eaten rather than fat or carbohydrate but, in practice, diets are mixed, for which heat production is equivalent to 6% of the energy value of the food. The origin of DIT still remains obscure, but it is suggested that it is partly from the digestive tract, caused by the digestive processes, and partly from the metabolism of absorbed food in the body.

Although it is not yet clearly demonstrated in humans, rat experiments have shown that DIT is increased by exercise. This means that less energy is available for the body as more is burnt off. The implications of this might be that those people with high DIT could eat more without gaining weight compared with those with low DIT.

Physical activity is the most variable factor affecting overall energy requirements. This will vary with (a) duration of activity, (b) body size (more energy required to move a large, compared with a small body) and (c) intensity of activity (such as the speed of walking).

Some of the **factors affecting protein requirements** are similar to those affecting energy requirements (Table 7.1). Of course, the poorer the quality of the protein the more will be required in the diet to satisfy nutritional requirements. This point is dealt with in Chapter 2. However, physical exercise does not increase protein requirement (unless body building has actually occurred and more muscle is laid down). This point is not often appreciated; thus, in a family, the man who does a hard day's physical work is given the lion's share of the meat on the assumption of his wife that he needs to be 'built up'. In fact, it is any children in the family who may need extra protein. The requirement for protein for infants and young children is much higher than for adults on a unit body weight basis. This is because of the extra requirement for growth.

If energy intake is low, then protein will first be utilised as an energy source, and secondly for growth and maintenance (its main function). There is thus a wastage of protein on diets inadequate in energy. An inadequate intake of energy can jeopardise the utilisation of protein for growth and maintenance to such an extent that signs of protein deficiency become evident (Section 8.5).

7.2 The adequacy of diet

Assessment of the nutritional adequacy of a diet is normally carried out by comparing its nutrient content (calculated from food intake, see Chapter 1) with the RDA. While such comparisons cannot be used as evidence that an individual is undernourished (this can only be confirmed with appropriate clinical examination and biochemical testing), it can be used as an indicator that a population may be 'at risk' of malnutrition.

Biochemical and clinical examination does show that, although there is very little overt malnutrition in the UK, anaemia (iron deficiency) in women is quite common. Other deficiencies which have been identified by these methods are summarised in Table 7.2. This table also shows that there are different deficiencies to be expected in the Third World.

Table 7.2 Summary of the main diseases of malnutrition with groups particularly affected

Nutrient	Groups mostly affected	Disease/ regions affected
Energy/protein	Infants, children	Kwashiorkor, marasmus T
Iron	Children, women, vegetarians	Anaemia W, T
Vitamin C*	Children, women	Anaemia W, T
Iodine	Adults	Goitre T
Vitamin D**	Children, women, vegetarians	Rickets, T osteomalacia T
Calcium	Pregnant & lactating women	Osteomalacia T
Vitamin A	Children	Xerophthalmia T (blindness)
Folic acid	Children, pregnant women	Anaemia W, T
Niacin	Maize eaters	Pellagra T

* vitamin C in the diet aids the absorption of iron
** vitamin D is produced in the skin by the action of sunlight; this is prevented by too much clothing
T, Third World; W, Western World

In addition to this clinical and biochemical evidence, dietary survey data from many parts of the western world show, for certain human groups, that some nutrients are supplied in the diet in inadequate levels with respect to RDA. Although in most dietary surveys, calcium, thiamin, riboflavin and vitamin C show averaged intake values above the RDA, values for individuals show that the lowest value of the range is well below the RDA. For iron, zinc and folic acid, even mean values may be well below the RDA. Obviously this means that there is no room for complacency, although some nutritionists argue that RDAs are set too high. It is clear that such data provide manufacturers with the ideal opportunity to market dietary supplements!

7.3 Pregnancy and lactation

Nutrition is the most important environmental influence in the development of the newborn. The developing baby gets all its nutrients from the mother's blood, so any factor which reduces that supply will reduce its growth. For example, severe malnutrition will lower blood pressure and this results in a small baby at term (end of 40 weeks).

During pregnancy there are increased demands for nutrients, as reflected in some increased RDAs (Appendix), in particular for folate, ascorbic acid and calcium. Supplementation of the diet is often recommended for folic acid (to prevent folic acid deficiency anaemia), iron and calcium. During pregnancy it is best to be careful about the use of high doses of nutritional supplements as high levels of vitamin A and D can be toxic and high levels of some of the water-soluble vitamins during pregnancy may lead to increased need in the infant after birth.

The demand for extra nutrients during lactation is higher than those of pregnancy. Thus, the RDAs for protein, some vitamins and iron are higher during lactation (Appendix). Although some energy has been stored by the mother as fat during pregnancy, which can be used at this time, there is a demand for increased energy intake. This is sometimes best met by increasing the number of meals per day and including cow's milk in the diet, which is a good source of many nutrients.

7.4 Infants and children

Human milk is a total food for the baby in the first few (up to six) months of life. Normally, not even water needs to be given to the infant in addition to milk. Milk provides all of the nutrients required until weaning, protects against infective agents, encourages a social bond between the mother and child and is low cost.

In the first five days of life of an infant, human milk is very different in composition and is called **colostrum**. This contains about half the immune factors which are passed from the mother to the infant (the remainder pass over before birth across the placenta). These factors are very important for the infant in combating disease. After about 10 days the composition of mature milk is achieved and remains at about the same composition until lactation is complete.

Infants that are breastfed have a lower mortality rate, even in the UK where there is less 'cot death' syndrome (sudden infant death) in breast-fed babies. In the Third World, the mortality rate among bottle-fed infants may be very high indeed and in such countries it is irresponsible to recommend bottle feeding except under exceptional circumstances. Human milk is low in bacteria and contains antimicrobial factors, so infants suffer fewer infections, in particular gastroenteritis.

Despite the advantages of breastfeeding, bottle feeding is still common in the UK, although more recently this trend has reversed as the medical profession is being a lot more supportive. Reasons for not breastfeeding have, in the past, included apathy of the medical profession, lactation failure and the fact that breastfeeding is no longer part of our cultural tradition. Infants that are not breastfed should be fed **formula** (also called **artificial** or **modified**) **milk** until at least six months of age. From six months onwards, all infants need extra energy and iron supplementation in addition to mother's milk.

It is important, if bottle feeding, that modified milk be used, and not just cow's milk, because human milk is very different from cow's milk. The gastrointestinal tract and the kidneys of the newborn are relatively immature at birth, and this will affect digestion and utilisation of food. The most important differences concern the fat, protein and minerals contents of milk. In fact, the balance of the major nutrients in cow's milk is completely different to that in human milk, including a much higher lactose (milk sugar) in the latter.

Table 7.3 Cow's milk compared with human milk

Component	Description	Comment
Fat	High in long-chain saturated fatty acids	Difficult to digest Inhibits calcium absorption
Protein	Three times higher	Infant kidneys cannot cope, excess amino acids can cause brain damage
	**Casein:whey protein ratio very high	Causes hard curd in stomach not easily digested
Minerals	Very high	Infant's immature organs cannot cope, leads to hypernatraemia*

*Hypernatraemia, is dehydration caused by a high level of minerals, especially sodium.
**Casein and whey protein are both milk proteins; there are more whey proteins in human milk

There is still a place, and probably always will be, for infant formulas, for women who cannot breastfeed their infants. In formulating milk, the aim should be to mimic human milk as closely as possible. However, any substitute for human milk can only be as good as current technology allows. Only in the present century have technological developments in dairy farming, food processing, paediatric nutrition, environmental hygiene and educational levels, combined with socioeconomic development, made widespread feeding of human babies with infant formula practical for the majority of people in the industrialised countries. Even so, it should be said that no infant formula exactly meets the infant's changing requirements during the first months of life.

The starting material for most infant formulas is cows' milk. It is necessary, in particular to reduce its protein concentration and reduce its mineral load, especially of sodium. Infant formulas can be made by mixing appropriate amounts of skim milk (to provide casein) and demineralised whey (by-product of the cheese-making industry, to provide whey proteins and lactose) to obtain a casein:whey protein ratio similar to that in human milk. The composition is completed by adding more lactose, minerals, vitamins and fat, and the product is homogenised and spray dried. Although it is only a crude imitation, it is nearer to the amino acid content of human milk and promotes good infant growth.

With all infant formulas it is essential that milk powders are made up correctly. Most manufacturers provide a scoop for the measurement of powder and instructions. Incorrect dilution can lead to:

over-concentrated formula: this can lead to obesity or even dehydration;
over-diluted formula: this is more likely to happen in the Third World, where a mother finds feeding the child by bottle too expensive and tries to economise by using less material.

Normally a baby will not accept solid food until about 4 months of age. Weaning commences when semi-solid foods are introduced into the diet and continues until after suckling has been discontinued. Weaning foods are first introduced as a sloppy gruel and then given to the child in a progressively more solid consistency as it gets older, finally replacing human milk altogether.

Weaning is a particularly vulnerable time for infants in the Third World. Weaning foods are usually prepared with starchy staples using non-potable water. Because of the water-holding characteristics of these ingredients the energy density of these foods may be as low as one-third that of human milk. The gruels fill up the child's small stomach capacity quickly, but provide little energy and can in some cases reduce intake of human milk. Feeding frequency may also be low if the mother is working. These combined factors lead to energy intake of the child below the RDA and protein energy malnutrition is a likely outcome (see Chapter 8). In addition, gruels can easily be contaminated with pathogens leading to severe diarrhoea.

7.5 The elderly

In old age many people, especially widowers living alone, pay very little attention to their diet. Vitamin C deficiency is quite common in this age group due to a lack of fresh fruit and vegetables in the diet. Generally, old people tend to eat less than the young, so it is important that the food they do eat contains adequate nutrients.

A big problem in old age is the decalcification of bones, called **osteoporosis**, which starts from middle age onwards and can leave the bones very prone to breakage after minor falls. The condition progresses more rapidly in women (after the menopause) than in men and is partly, but not entirely, diet-related. Low vitamin D status has been shown to occur in some old people and this would lead to loss of body calcium. (Vitamin D levels can be raised by exposure of the skin to sunlight which allows the body to make its own supply.) However, in many people osteoporosis cannot be prevented only by a high intake of either calcium or vitamin D, although it is important to ensure that the intakes of these nutrients is adequate. Physical activity in particular is very important for reducing the rate of decalcification.

7.6 Vegetarians

In less-developed countries, vegetables often provide the bulk of the diet, not necessarily through choice but sheer economics. It is possible to obtain nearly all nutrients required for health from vegetable sources, although more care is needed with such a diet to avoid nutrient deficiency. Table 7.4 shows the main nutrients provided by the various classes of vegetables.

Table 7.4 Good sources of selected nutrients* provided by four plant food groups

Food group	Nutrient
Cereals	Energy, protein, dietary fibre**, vitamin B complex**, iron**, calcium**
Legumes, oilseeds, nuts	Energy, protein, dietary fibre**, iron**, calcium**, vitamin B complex**
Roots	Energy, protein, some vitamin C
Fruit and vegetables	Vitamin C, vitamin A, iron, calcium, vitamin B complex, dietary fibre

* nutrients selected may be low in the diets of vulnerable groups
** particularly high in the hull

There are various groups of vegetarians. Those that only eat food of vegetable origin are **vegans**, while others in addition eat eggs and milk products (**ovolacto-**) or even fish (**pesco-**). Those that eat some animal products have little difficulty obtaining all their nutrients from their food, but vegans have a problem with vitamins B_{12} and D, neither of which are present in plant products. There should not be a problem if vegans make sure that they get adequate exposure to sunlight to make their own vitamin D and most vegans take vitamin B_{12} supplements. (Vitamin B_{12} is made by microorganisms and concentrates in the marine and terrestrial food chains, being stored in the liver of animals, which is a very rich source.)

In addition to these two vitamins, it is important that vegetarians obtain adequate iron and calcium in their diets. Calcium is relatively low in plant foods, except the hull of grains and legumes, while iron, as well as being quite low, is not so well absorbed from plant foods. Often vegetarian diets are high in dietary fibre which tends to bind minerals and render them unavailable.

7.7 The sick, institutionalised and those with specific disorders

The main thing to look for in institutional catering is that the level of vitamin C in the diet is adequate. Often foods are maintained at high temperatures for a long time between cooking and consumption and this greatly reduces the vitamin C content. After surgery, requirements for nutrients, especially vitamin C may increase.

7.8 Diabetics

Biochemical and physiological aspects of diabetes (mellitus) are explained in Section 5.6. With both types of diabetes, juvenile and mature-onset, the main aim of management is to maintain blood glucose levels within as narrow a range, and as near to normal levels, as possible. The closer the control, the better the chances of preventing common complications of diabetes involving small blood vessels (leading to kidney disease, retinal damage and loss of eyesight). At the same time, a prudent diet should be followed to minimise damage to large blood vessels caused by atherosclerosis, leading to coronary heart disease and restriction of blood supply particularly to the legs.

Very often mature-onset diabetes is associated with being overweight and then blood glucose can be controlled within normal levels merely by weight loss. Apart from weight control, the advice on diet for both types of diabetics has changed recently. Until recently, a high fat, low carbohydrate diet was considered to be the best for diabetics, but this regime led to a high risk of diabetics developing heart disease.

Doctors working in Oxford have compared a low carbohydrate, high fat diet, for both types of diabetics, with a high carbohydrate, high fibre diet. Although wholegrain cereals are helpful in controlling glucose levels, legumes (beans) give better control. In addition, diets high in wholegrain cereals and beans help to reduce the amount of insulin required for insulin-dependent diabetics. Following clinical trials of this nature, diabetics are now recommended to eat diets similar to the dietary guidelines for the general public (Section 1.4), with perhaps more emphasis on increasing the dietary fibre intake, especially from legumes and cereals.

8 Malnutrition: over- and undernutrition

Malnutrition is nutrition which deviates from normal. If deviations are large then clinical signs will appear. Malnutrition can be divided into primary and secondary malnutrition. The former concerns a lack or excess of energy and/or one or more nutrients in the diet. Secondary malnutrition results from a pre-existing disease condition, such as uncontrolled diabetes or cancer. In this chapter some of the major issues of primary malnutrition facing poor and wealthy countries are outlined. It is one of the ongoing ironies of modern life that while we see television programmes about starving populations in Africa and other disadvantaged societies, one of the most serious health problems of modern industrialised communities is coronary heart disease which is associated with obesity caused by excessive food intake.

8.1 Overnutrition and obesity

Body fat acts as a store of energy, protects internal organs from injury and insulates the body, thus helping to conserve heat. However, too much body fat can have a detrimental effect on health. Indeed, various assurance companies have worked out the ideal weight for longevity for people of various heights (Figure 8.1). The greater the weight for height is from normal the more the obesity and the shorter the life expectancy. Obesity can be assessed using the **body mass index** (BMI) which is defined as W/H^2, where W is weight in kg and H is height in metres. The best range for health lies between 20 and 25 for men and women.

A very high proportion of the UK population are overweight and the proportion increases with age, with over half the men and women in the 60–65 age range being overweight. Also 14% of the population are so overweight as to be called **obese** (BMI > 27.5). People who are overweight have increased risk of:

* coronary heart disease
* gallbladder disease
* hypertension (high blood pressure)
* diabetes
* some cancers (endometrium, gallbladder)
* death during surgery (thick layers of fat are difficult to cut through without causing bleeding and make organs difficult to identify)

Weight increases beyond the ideal because energy intake exceeds energy expenditure and the excess is laid down as fat, as it cannot be lost from the normal healthy body. It is not necessary to assume that a person who is overweight eats more food than a lean person, as everyone's energy require-

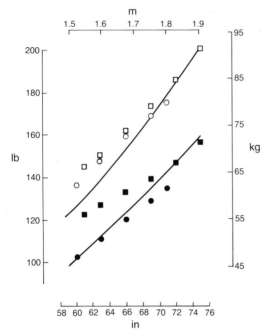

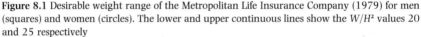

Figure 8.1 Desirable weight range of the Metropolitan Life Insurance Company (1979) for men (squares) and women (circles). The lower and upper continuous lines show the W/H^2 values 20 and 25 respectively

ment is different. Often overweight people eat less than lean people of the same height and sex, but they have lower energy requirements for their basal metabolism.

These days sedentary occupations lower the demand for energy intake and the adjustments to diet may not be made accordingly. A wide variety of tempting foods available on supermarket shelves may have something to do with this. Certainly it is possible to get rats to overeat and become fat by providing them with a diet of varied human snack foods, despite the fact they would remain lean all their lives on plain 'laboratory chow'. Food intake is governed by a whole range of factors, including psychological ones – some people eat more when they are stressed or bored, while others eat less.

To lose weight it is necessary to reduce energy intake or to increase energy output. A combination of a reducing diet and increased exercise is the best approach. The amount of energy expended on even quite strenuous exercise is disappointingly low. However, regular exercise tones the body and may stimulate greater heat production in response to food intake (diet-induced thermogenesis, see Section 7.1), which may waste some of the excess energy intake. Regular exercise also contributes to the discipline necessary to maintain a low energy intake over a long time.

Slimming diets to produce moderately rapid weight loss would be in the range 800–1500 kcal (3300–6300 kJ) per day, depending on energy requirements, although some **very low calorie diets** have come onto the market in recent years which have less than 600 kcal calories (2520 kJ) per day. These are specially formulated products which have added nutrients to make sure that the reduced energy intake does not jeopardise adequate intakes of other nutrients.

8.2 Coronary heart disease

Coronary heart disease (CHD) is one of the greatest causes of death in Britain and many industrialised countries. It is less common in rural communities in developing countries, although increasing among the well-to-do. Forty years ago CHD was a disease of men of the affluent classes in Britain, but the years 1950–65 saw involvement of all social classes. Of particular concern is the marked increase of this 'modern epidemic' disease in the age group 35–44 years. The incidence of CHD has now reached a maximum in the UK, but as yet has shown little sign of tailing off except among the professional classes. Until recently, Finland, the USA and Australia topped the list of the world's highest incidence of CHD, but in all these countries the rate has declined markedly, leaving Britain at the top. This decline is thought to be due to increasing public awareness of the role of diet in the development of CHD, as well as other factors such as less smoking. The relationship of diet to the prevention and treatment of CHD has been the subject of much study and still remains contentious, despite the fact that one of the principal methods of treatment of the disease is by dietary restriction or modification.

CHD arises from the failure of the coronary arteries to supply sufficient blood to the heart tissue. In most cases the disease is associated with **atherosclerosis** (see below) of the coronary arteries, which supply the heart tissue with blood. Symptoms include **angina**, which is pain in the chest, caused by exercise. Patients with this must curtail sudden exercise. A heart attack often occurs because of the formation of a **thrombus** (blood clot) in an atherosclerotic coronary artery which blocks it and cuts off the blood supply. Part of the heart muscle may die (necrosis) and be replaced by scar tissue. Survival depends on the extent of the damage.

Table 8.1 Some risk factors in coronary heart disease

Maleness
Age
Family history of coronary heart disease
High blood cholesterol level
High blood pressure
Diabetes
Obesity
Cigarette smoking
Certain diets (e.g. high in saturated fats)
Lack of exercise
Stress
Soft water consumption

Some of the risk factors for CHD are listed in Table 8.1. These have been identified from **epidemiology** (study of the disease patterns within and between communities). As mentioned in section 6.9 this is an inexact science, and can never be used to prove a relationship between the incidence of a disease and an aspect of lifestyle as there are so many factors which differ between communities. However, it can be used to indicate possible areas where relationships exist so that further studies can be undertaken (in nutrition these would fall into the broad categories of animal experimentation or clinical studies). Many epidemiological studies have now been undertaken or are underway on CHD and some have shown three primary risk factors associated with it: high blood cholesterol level, high blood pressure and cigarette smoking.

The pathological basis of CHD is atherosclerosis in one or both coronary arteries. Human beings are remarkably prone to atherosclerosis, which starts off in childhood as fatty streaks containing cholesterol in the lining of the major arteries. Depending on diet, these may develop into plaques or even complicated lesions. If the lining of the artery is disrupted, then blood platelets may be attracted to the surface and initiate a blood clot.

There are several hypotheses for the development of CHD and the underlying atherosclerosis. The most favoured one is the lipid hypothesis which is shown in Figure 8.2.

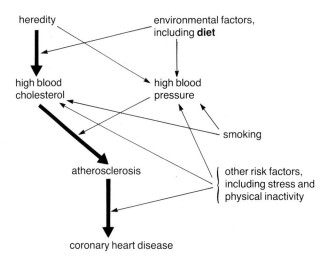

Figure 8.2 The lipid hypothesis for the development of coronary heart disease

8.3 Effect of diet on plasma lipids

There is not a lot of direct evidence, apart from that from epidemiology, to link diet with CHD. For this reason, there is, and will continue to be, controversy over the precise role of diet. However, there are a lot of data from animal and clinical studies to show that diet does influence two of the primary risk factors in CHD: the levels of blood cholesterol and blood pressure. The influence of various dietary factors are shown in the Table 8.2.

Table 8.2 Dietary factors influencing blood pressure and cholesterol levels

	Blood pressure	*Blood cholesterol*
Raising	Sodium (salt)+	Saturated fat**
		Dietary cholesterol*
Lowering	Potassium?	Soluble dietary fibre*
	Calcium?	Plant protein?
		PUFA**
No effect		Monounsaturated fatty acids
		e.g. oleic acid in olive oil
		starch
		sucrose
		alcohol

*, small change; **, large change; ?, some evidence, but research still needs to be done to establish; +, only some people affected; PUFA, polyunsaturated fatty acids.

Foods which may contain a number of blood cholesterol lowering factors are particularly effective. Thus peas and beans show marked blood cholesterol reducing activity, which could be due to the lack of cholesterol in them, the fact that the protein is of plant origin or the high dietary fibre content. Generous intakes of alcohol may lead to high triacylglycerol levels in the blood, but the greater risk factor, blood cholesterol level, is not affected. However, moderate alcohol intakes can lead to a higher level of **high density lipoproteins** (HDLs) in the blood, These are one of the fat-transporting systems in the blood and a high level is thought to be advantageous in protecting from CHD. Contrary to pronouncements in the press and elsewhere, sucrose has no effect on blood cholesterol level, even at high intakes.

Western diets are high in sodium and low in potassium, while the reverse would be true on a strict vegetarian diet. The relationship between salt intake and high blood pressure is still subject to debate; it appears that only some people respond to high salt diets by increased blood pressure. However, the rise of blood pressure with age seen in people living in western societies does not mean that this is inevitable in humans. Primitive peoples (such as some people of the New Guinean islands) living on low salt diets show no rise as age advances. On the other hand, the Japanese have very high blood pressures, one of the highest rates of stroke in the world and a very high salt intake (twice as high as the British).

Prevention of CHD should begin in early childhood, with emphasis on the avoidance of high blood cholesterol levels. The COMA 1984 Report published by the DHSS on *Diet and cardiovascular disease* made some general recommendations on how to modify the British diet to reduce the incidence of CHD and these recommendations were outlined in Table 1.3. For people who already have atherosclerosis then some regression might be possible by modifying the diet, as this has been shown to occur in rhesus monkeys. However, for advanced calcified lesions one would not expect much regression, so the other way of preventing CHD would be to prevent thrombosis. Polyunsaturated fats will inhibit thrombosis, particularly if they contain Ω-3 fatty acids (see Figure 2.14), which are found in fish oils.

8.4 Undernutrition

Undernutrition can be general (starvation) or specific (lack of a certain mineral or vitamin). On a worldwide scale there are some types of deficiency syndromes which dominate. These are **protein energy malnutrition** (PEM: lack of dietary energy and protein), **xerophthalmia** (dry eye, caused by a lack of vitamin A), **goitre** (swelling of the thyroid gland in the neck due to lack of iodine) and **anaemia** (lack of haemoglobin in the blood and therefore reduced ability to transport oxygen round the body due to lack of iron). These diseases of malnutrition are discussed in this chapter. Although other deficiency disorders occur, such as pellagra (lack of niacin), rickets (lack of vitamin D) and folate deficiency anaemia, they are not so widespread.

8.5 Protein energy malnutrition

Kwashiorkor and **marasmus** are the extreme manifestations of PEM, although cases showing symptoms in between the two will be found commonly. The typical appearance of these diseases can be seen in Figure 8.3.

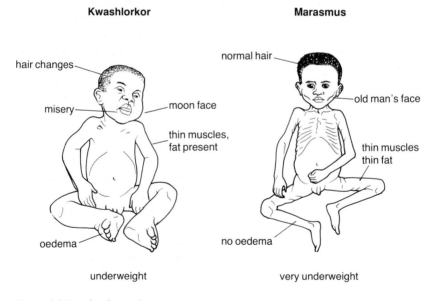

Figure 8.3 Kwashiorkor and marasmus

The main clinical signs of kwashiorkor are swelling of the legs (oedema), sparse hair, 'flaky paint' appearance of the skin and pigmentation changes of the hair. The 'moon' face shows very little interest in surroundings. Biochemical changes are profound, causing accumulation of fat in the liver which can lead to cirrhosis (permanent damage of liver caused by infiltration with scar tissue). There is also marked lowering of blood albumin level which is responsible for the oedema.

The child with marasmus has a low weight for its age and has thin arms and legs: a 'wizened' appearance. There are few biochemical changes and in some respects a child developing this condition in response to a poor diet has adapted better than one developing kwashiorkor. In any case, children with either condition are in grave danger and any feeding should be carried out with care: they should be given frequent servings of very small amounts of food, as digestion and absorption is reduced. The reason is that PEM causes pancreatic and intestinal cells to atrophy (die). Thus less digestive enzymes are produced and the absorptive surface is greatly reduced.

In the 1960s kwashiorkor was ascribed to a protein deficiency, while marasmus was considered to be caused by a lack of energy in the diet. It is not usual these days to regard the two diseases as so separate. In the 1960s kwashiorkor was regarded as the predominating PEM disease in Third World countries, so high protein supplements were promoted through aid agencies to cover the 'protein' gap. We now know that marasmus is equally, if not more common. In the early 1970s nutritionists started to rethink the relationship between protein and energy in the diet. This resulted in the publication by the Food and Agriculture Organisation of the United Nations in 1973 of the *Energy and protein requirements*, which gave lower RDAs (recommended daily amounts) for protein than had been previously recommended. At about the same time, in India, it was observed that if two similar children, matched for age and sex, were eating the same poor diet, one would develop marasmus and the other kwashiorkor, despite the fact that both diets provided the same amount of protein.

Dietary survey data from poor communities within various developing countries were also very revealing. They showed that in all cases energy intake was much lower than RDA. This is important, as lack of energy in the diet jeopardises the use of protein for growth and maintenance. If the energy intake is low, then protein is used as an energy source and not for its primary function of growth and maintenance. However, as these latter requirements are not satisfied, the child shows symptoms of protein deficiency (kwashiorkor) although it is really energy intake which is deficient.

These days most aid programmes supplying food aid, aim to increase the intake of energy in the diet. To do this it is important to take account of the acceptability of foods, as people will not eat their energy requirements if they find the food unpalatable, unfamiliar or monotonous. If energy intake is satisified, then most staples in the diet, like wheat or rice, will provide enough protein. The main exceptions to this are the tropical crops cassava (a root crop), plantain (a type of starchy banana) and sweet potato, which are very low in protein and care should be taken that they are supplemented with other protein foods.

Anorexia nervosa is analogous to marasmus in clinical appearance, but, in contrast, it is usually encountered in developed societies among the well-to-do. It mainly affects teenage girls who have an obsession with weight loss, which becomes excessive. The whole body becomes wasted, including muscle tissue, and sexual development is stopped. The girl has an abnormal sense of body image, seeing herself as fat, even in an emaciated state. She is preoccupied with diet, exercise and death, and becomes a great cause of concern to family and friends. Treatment can be by tube feeding in early stages. Rehabilitation involves accepting a realistic weight goal and, often, making the psychological adjustments needed to develop from childhood to adult life. For a further discussion of anorexia nervosa see Cornwell & Cornwell, *Drugs, alcohol and mental health* in this Social Biology Series.

To avoid malnutrition it is important to eat a wide variety of foods, especially if the diet is largely of vegetable origin. Table 8.4 summarises the value of mixed diets so often advocated by nutritionists for health and well-being.

Table 8.4 Value of mixed diets

---

(1) Improves protein quality by improving amino acid pattern.

(2) Increases food acceptability: monotonous diets lead to energy intakes lower than requirements. This means protein is used as an energy source, increasing the protein requirements.

(3) Greater variety of foods gives a better balance of vitamins and minerals.

(4) A variety of foods has a dilution effect on toxins present in individual foods.

NB *The points above are based on the concept of a 'balanced' diet, with emphasis on avoiding nutrient deficiency. In addition, more recently, dietary advice for 'healthy' eating is to also consider the proportion of the major food components, especially fat (Chapter 1).*

8.6 Goitre

Iodine is required for the production of thyroid hormones from the thyroid gland in the neck. These hormones regulate several systems in the body, including basal metabolic rate (Chapter 6), growth and reproduction. If dietary iodine is deficient, then the thyroid gland enlarges to form a goitre. In a small number of cases low iodine in the diet may be exacerbated by the presence of **goitrogens** in some vegetables, which prevent the thyroid gland taking up iodine. The cabbage family of vegetables contains goitrogens, but in normal quantities in the diet of a well-nourished person this causes no problem. In adults, goitre causes fatigue and weight gain, but there is a particular danger for pregnant women with goitre. The child can be affected in the uterus, and will be born a **cretin** with low IQ and physical retardation.

Seafood is high in iodine, as are vegetables grown in iodine-rich soil. Manufacturers normally add iodine to table salt and, in this way, goitre has been eliminated in many parts of the world, although it is still a problem affecting some 200 million people worldwide in more inaccessible areas.

8.7 Xerophthalmia

Vitamin A has a number of physiological functions in the body. If a person is deficient in vitamin A, then the retina contains less **rhodopsin**, the visual pigment needed for seeing at low light intensities. Vitamin A is also important for the maintenance of healthy **epithelial** tissue. Epithelial tissue is a protective coating found on external surfaces, such as skin, and internal ones, usually mucous membranes. Both of these can become keratinised (invaded with dry hard protein) in vitamin A deficiency and this usually leads to infection. The first effects of vitamin A deficiency are **follicular hyperkeratosis** (keratin plugs in hair follicles causing a roughened skin surface), then loss of dark adaptation.

The symptoms already mentioned are reversible with vitamin A refeeding. However, prolonged deficiency of vitamin A can lead to permanent blindness, that is xerophthalmia (dry eye), which is common in many parts of the Third World, including SE Asia and Africa. As much as one-third of the children may be affected in these areas. The most common onset is from 2–5 years, and it is often associated with kwashiorkor. It is estimated that 20 000 children are blinded every year for this reason alone. Other diseases may be associated with it, especially diarrhoea, measles and other infections.

Well-nourished people usually have adequate stores of vitamin A in the liver to last months, or sometimes years, before the first signs become evident. Nevertheless, in the developing world whole populations may be deficient. In these countries, very little pre-formed retinol is eaten, as it is only present as such in animal products which are too expensive. Also people do not eat enough green vegetables containing carotenes which can be converted to vitamin A by the body.

Xerophthalmia is a form of preventable blindness. Early xerophthalmia can be treated by corneal transplants, but obviously this is not the long-term answer. In developing countries, what is needed is a long-term policy. People should be taught to eat more fresh green vegetables and to cook with palm oil where it is available. Xerophthalmia often develops through ignorance. In areas where there is a severe problem it is quicker to give synthetic vitamin A. This can be administered by fortifying a common foodstuff such as sugar. In India, massive doses (60 000–90 000 μg retinol) at six-monthly intervals have been tried with success.

8.8 Anaemia

There are a number of anaemias including pernicious (lack of ability to absorb vitamin B_{12}, or lack of that vitamin in the food, such as in some vegetarians) anaemia. Lack of folate can also cause anaemia. However, the type of anaemia which is most prevalent worldwide is **iron deficiency anaemia** and this is discussed here.

Iron is not very readily absorbed by the body, and most in the diet is passed out in the faeces. The best form of iron for absorption is haem iron, present in haemoglobin in the blood and high in red meats. But there are other dietary factors which increase iron absorption: the most important is vitamin C. The mechanism responsible is thought to be that iron in the diet in the ferric form is changed to the ferrous form by vitamin C and is more easily absorbed as such. Alcohol also increases iron absorption and this is well known in countries like France, where a lot of wine is drunk. It has been found that wine drinking can lead to excess iron being deposited in the body, although this does not have any serious consequences.

Unfortunately, there are also substances in food which inhibit iron absorption and therefore lower its **bioavailability. Phytate** and excess phosphates can lead to the formation of insoluble iron salts in the gut, which are passed out in the faeces without absorption. Phytate is present in seed products such as wholegrain cereals, peas and beans. Dietary fibre can also bind iron and prevent its absorption. All these factors are present in plant foods, so it is not surprising that those eating vegetarian diets are more susceptible to anaemia, particularly as they do not have access to haem iron.

9 Food from production to consumption

9.1 Introduction

The wide variety of food in the supermarkets in the western world reflects the activities of sophisticated food industries. With over 70% of the food eaten coming through the food industry in a country like the UK, the public are naturally concerned about the effects that processing has on the food. The more highly developed the food industry, the greater is the responsibility for feeding large numbers of people with safe food and providing them with all their nutrient requirements. Trained personnel are essential in food manufacture (see Plate 2). Most developing countries also have embryo food industries, which are growing as population pressures in large urban communities increase demand for readily available food, which is labour-saving for working women. Of course, processing is not just confined to the food industry, but also applies to any preparation or cooking carried out at a domestic level. Although much is documented on the effects of industrial processes of all types on the nutritional and microbiological value of food, very little is known about the effects of domestic processing in this respect.

In industrialised countries there is a great deal of public criticism concerning activities of the food industry, some of which has been justified, but much of which is unfounded. Criticism more recently has centred on (*a*) too many additives used, (*b*) too much profit made and (*c*) selling food of dubious nutritional value. Most recently consumers have been worried further by reports of microbiological contamination of food: most notably *Salmonella* in eggs and poultry and *Listeria* in cooked, chilled convenience foods and soft cheeses. In the Third World it is the multinationals who come in for most criticism and this revolves around the nature of the products manufactured; for example soft drinks (providing energy only as 'empty calories', that is containing very few other nutrients), beer (encouraging the misdirection of the family funds) and infant formulas for babies (most of the criticism has been aimed at advertising campaigns, fortunately less frequent more recently, where mothers in poor areas of the world were encouraged to bottle feed their babies rather than breastfeed, a practice leading to malnutrition and death in many cases). However, in the Third World the food industry is contributing to development in less-publicised areas: cereal milling, sale and distribution of dried fish and canning for local consumption are examples.

The last five or ten years has seen very big changes in the attitude of the food manufacturer towards nutritional issues in the western world. It has coincided with greater emphasis by governments on the link between diet and health (see Chapter 1 for guidelines on 'healthy' eating). These issues have been promoted

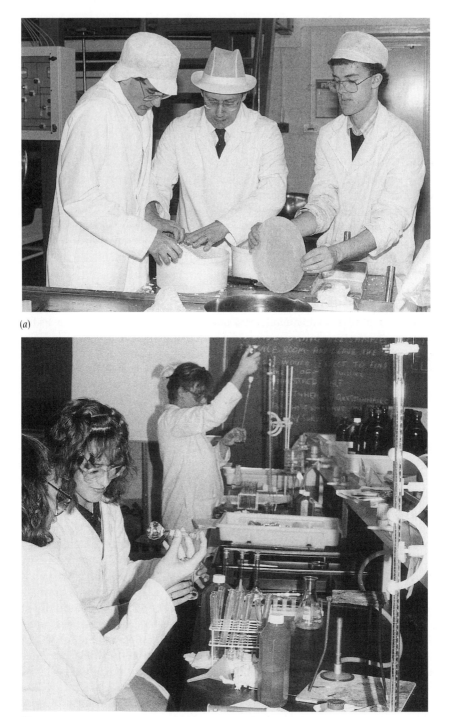

(a)

(b)

Plate 2 Training food scientists for the food industry; (a) cheese-making, (b) examining the nutritional quality of food (photographs courtesy of University of Reading)

by consumer pressure via various activist groups such as consumer associations. Recently, food manufacturers have taken a more open approach to nutrition rather than the previously-held defensive attitude when nutrition was a thing not mentioned in a food factory. The result of this has been a greater emphasis on nutrition labelling on products, new product development and the manufacture of 'healthy' alternative products. The aim is to encourage consumer choice, with the realisation that it is, after all, the consumer who decides on diet composition given that the appropriate foods are available.

Although the reasons for humans first processing food are lost in obscurity, it is likely that the use of fire improved the flavour and texture of foods. While changes in such sensory characteristics are important, processing, especially heating, reduces the danger of pathogenic microorganisms being present and therefore the incidence of disease. This and other reasons for processing foods are listed below.

Reasons for processing food

* Storage, preservation
* Prevention of disease
* Improvement of digestibility and nutritive value
* Palatability/acceptability
* Fortification
* Destruction/reducton of toxic components
* To change the physical properties of food

9.2 Health hazards from food

Microorganisms

Two main types of organisms are of concern to those manufacturing or providing food: **spoilage organisms**, and **pathogenic organisms** causing food-borne diseases. Spoilage organisms are mainly yeast, moulds (fungi) and bacteria. The presence of these organisms in food provides the need for preservation techniques. One main preservation technique is to remove water by freezing, salting or drying; these remove the water available to the organism and stop its growth.

Food poisoning is a worldwide phenomenon. In industrialised countries it is chiefly caused by bacteria. Bacteria are divided into three main groups: **psychrophiles, mesophiles** and **thermophiles**. This grouping indicates the temperature range in which they will grow optimally, that is 15–18 °C, 30–45 °C and 55–75 °C respectively. Pathogens are mesophiles and the further the temperature of the environment is from the optimum growth temperature, the more slowly they will grow, a fact which forms the basis of some preservation techniques. Some mesophiles are **psychrotrophic**, that is they can grow at refrigeration temperatures even though their optimum growth temperature may be between 25 and 40 °C. Unfortunately, this group includes three food-poisoning bacteria: *Listeria monocytogenes*, *Yersinia enterocolitica* and *Clostridium botulinum* type E. Water, as well as food, carries food-borne disease, particularly in less-developed countries where supplies of potable water are scarce.

With increasing centralisation of the food industry, problems of parasitic infection have been largely tackled through better hygiene and meat inspection, but bacterial pathogens still present a problem, even in industrialised countries. There are relatively few bacterial species which can cause food poisoning. These include *Salmonella, Staphylococcus, Clostridium* spp. (especially *Clostridium botulinum*). Some of these bacteria produce an infection within the body, while others produce toxins in the food which are a hazard when eaten. Examples of infectious organisms are *Salmonella* spp. which are common in farm animals. Most poultry carcasses are contaminated with these organisms. In animals these pathogens may cause mild diarrhoea or be completely asymptomatic.

Among the toxin producers are certain strains of *Staphylococcus aureus* which produce a toxin when growing in food. One in every four people is a carrier of this type of organism and these people are a particular hazard in food manufacture and catering as they can contaminate food. Some food retailers are particularly keen that employees should not be *Staph.* carriers if they are involved in the preparation of fresh foods, such as sandwiches, which are to be eaten without further cooking. *Staph.* carriers can be identified from the microbiological culture of nasal swabs.

Although each year in England and Wales there are some 20 000 reported cases of food poisoning, this only represents a very small proportion of the total. It is estimated that only about one in eight cases of gastroenteritis is reported. Sources of infection are through bad handling and contamination from organisms resident in the nose and bowel via hands, or by direct contact between raw and cooked meat from hands or cutting equipment. To avoid the latter, facilities should be installed in all outlets for cutting and handling raw meat separately from cooked meats. Hands should not be allowed to touch meat, paper or plastic film should be used instead.

The number of organisms required (the infective dose) to cause gastroenteritis is relatively small for some infections like dysentery, although this will always depend upon the person and their susceptibility to infection (for example well-nourished people are less susceptible than malnourished ones, and babies and older people may be more at risk). It is possible that an infection can be severe enough to cause a general fever (enteric fever) with the organisms gaining access to the body and not just confined to the gut. There is a danger that those who get such a fever, even though they recover from the illness, never lose the organism and become carriers for life. It is very common to find asymptomatic typhoid carriers in many parts of South America for example. This was the origin of the 1964 Aberdeen typhoid outbreak caused by a large can of corned beef imported from Argentina. The can had been produced in a factory downstream from a town estimated to have about 200 carriers of typhoid (*Salmonella typhi*). Raw river water was used to cool the cans after processing. During cooling it is well known that can seams may momentarily open and introduce cooling water!

Although infections and intoxications both produce gastroenteritis characterised by diarrhoea and vomiting, bacteria which produce infections produce a different pattern of illness from those that cause intoxication. Intoxication is, in general, much more rapid in onset, while infections may last longer. Table 9.1 indicates some of the more common bacterial species causing food poisoning.

Table 9.1 Food-borne bacterial intoxications and infections

Organisms	Food with which associated	Other comments
Intoxications		
Bacillus cereus	Any foodstuff, esp. cereals	Chinese take-away food implicated in past.
Clostridium botulinum	Canned food and fermented fish	Fish products particularly in Japan (type E): need for prompt action with antiserum. Toxin is heat-labile
Clostridium perfringens	Deep stews in institutional catering	Toxin not very heat-labile.
Staphylococcus aureus	Precooked food handled by carriers	Toxin is quite heat-stable.
Infections		
Salmonella typhi	Water or shellfish contaminated with sewage Contaminated food	Humans are the main reservoir for this organism. After typhoid fever people can become carriers for life.
Salmonella spp.	Animal products, esp. poultry	Common in animal faeces.
Yersinia enterocolitica*	Raw milk, meat, but esp. pork products	Environmental organism.
Campylobacter jejuni*	Animal products, esp. poultry	Common in animal faeces.
Listeria monocytogenes	Cook/chill meals, soft cheeses made from unpasteurised milk	Environmental organism, may be normal commensal in human gut. Can (rarely) cause fever leading to abortion and meningitis.

*, also produces toxin, but infection considered most important.

Natural toxins in food

There are many biologically active substances in foods, many of which have not yet been identified. Although certain foods react adversely with some people, leading to headaches, nausea, indigestion or other symptoms, these are often individual responses, such as allergies, not shown by the population in general. On the whole, most foods we eat do not cause a toxic response because over the centuries humankind has learnt to avoid eating toxic substances by:

* selection of food crop varieties;
* eating a mixed diet;
* processing, especially heating.

The use of fire later in human evolution introduced a whole new range of foods into the diet, as the action of heat destroyed many toxic substances. Although there is a long list of substances known to be toxic in foods, many are only mildly so. Indeed, only a few of them pose a hazard to health (see Table 9.2).

Table 9.2 Natural toxins of plants of particular public health significance

Toxin	Food affected	Primary disease/symptom	Detoxication	Comments
Aflatoxin	Peanuts	Liver cancer	Remove from diet	Ensure low levels by good QC
Cyanogens	Cassava Fruit kernels, e.g. bitter almond	Annoxia, can be fatal	Crush or grate cassava wash in running water, cook in open vessel Discard bitter varieties	
Favism	Vicia faba (faba or broad beans)	Haemolytic anaemia of certain persons; can be fatal	Remove from diet	Inherited enzyme deficiency especially Med. region
Haemag-glutinin	Red kidney beans	Acute gastroenteritis	Cook well by boiling for at least 10 minutes	

Med., Mediterranean; QC, quality control.

Aflatoxin presents a particular difficulty for tropical countries, because it is only under conditions of high temperature and humidity that the ubiquitous spoilage organism *Aspergillus flavus* (a mould) produces aflatoxin, a chemical substance, which is now known to be a very powerful **carcinogen** (substance causing cancer). Aflatoxin was first identified in the 1960s in a consignment of groundnut (peanut) meal imported into the UK from Brazil for animal feed. Although aflatoxin can be found in other foods in the tropics, in international trade it is its presence in peanuts which is important. These days any exporting country has to monitor the aflatoxin level of peanuts very carefully. If quality control is not adequate then shipments will be refused by the importing country. Unfortunately, good quality control does not always extend to produce intended for home consumption. In some West African countries the high incidence of cancer of the liver is thought to relate to high intakes of aflatoxin in the diet.

Cyanogens (cyanide-producing substances) are found mainly in bitter al-monds and cassava (a tropical root crop widely eaten as a dietary staple) and older, black varieties of Lima bean (butter beans). In the literature, there are reports in the past of outbreaks of poisonings of humans resulting in a number of deaths in Mauritius and Puerto Rico due to black butter beans, but this variety is not now used in international trade. However, even to this day, from time to time, there are reports of deaths in Nigeria due to isolated cases of people eating a very heavy meal of cassava.

For cyanide to be formed, the cyanogen has to come into contact with an enzyme also present in the plant. While the cyanogen is situated inside the plant cells, the enzyme is outside and the two will only come into contact due to damage of the cell structure by cutting or bruising the tissue. To eat cassava safely there are several things that can be done. The sweet varieties should be selected for consumption, as the bitter varieties, on the whole, have a much higher level of cyanogen than the sweet varieties. All varieties should be subject to special preparation before eating. The cyanogen should be converted to cyanide, which can then be released by washing (the root is often placed in running water) or heat (cyanide is highly volatile). Conversion to cyanide can be achieved by deliberately damaging the cells and letting the enzyme come into contact with the substrate, by grating, or pounding the root. Finally, cassava should be cooked in plenty of water with no lid on the pot, to allow the cyanide gas to escape (otherwise it will condense on the inside of the lid and fall back into the pot). After cooking the water should be discarded.

Favism is an individual response of some people to the toxic favism factors, **vicine** and **convicine**, present in broad beans (*Vicia faba*). Susceptible people are mainly from the Mediterranean region (such as the islands of Sardinia and Rhodes, where about 5 in 1000 people may be affected). Favism occurs because of a deficiency (not total lack) of a certain enzyme in the body. Despite this deficiency, people with favism show no abnormality until they eat broad beans, when the lack of the enzyme allows the favism factors to intoxicate the body. The red blood cells break down causing **haemolytic anaemia**, which can be fatal, especially in children.

Favism factors are quite stable to processing, including heat treatment, and, for this reason, susceptible people should avoid eating broad beans or, in some cases, even avoid inhaling the pollen!

Haemagglutinins (or lectins) are proteins which are found in many seeds and plant foods, but the main one of concern to us is that from the red kidney bean. This haemagglutinin is resistant to digestion and, if the beans have not been cooked sufficiently, they will pass through the entire gut undigested and can be detected in the faeces. In the small intestine they can attach themselves to the gut wall and, in small amounts, change gut motility (movement), resulting in vomiting and diarrhoea. In large amounts the gut wall can be disrupted to the extent that gut bacteria can invade the body tissues and death can result.

In the UK between 1976 and 1979, when there was a transitory fashion for eating soaked, raw kidney beans as part of a 'whole-food' fad, this led to several cases of severe gastroenteritis resulting in hospitalisation for some, but fortunately no deaths. The introduction of new food habits in the UK with increasing interest in preparing 'chile con carne', a Mexican dish with red kidney beans, has increased the requirement to be aware of this health hazard. Red kidney beans are quite safe if they have been properly boiled for about 10 minutes, as this denatures the haemagglutinin and renders it inactive.

Food allergies are not considered public health problems, as they only affect certain hypersensitive people. Nevertheless, they occur sufficiently frequently to cause much consumer concern. A food allergy is an **abnormal immuno-logical response** to a food and would lead to a raised level of immunoglobulins

(IgE) in the blood. Certain proteins in the food act as **antigens** and if they enter the body elicit the formation of **antibodies** (IgE). If there is no involvement of the immune system, a syndrome cannot be strictly called an allergy, although, unfortunately, some medical practitioners are using the term to describe any kind of **food intolerance**, including those not involving the immune system.

Food allergies are difficult to diagnose because the tests are not ideal. The usual one used is a crude skin scratch through an extract of the food. Another difficulty in diagnosing allergy is that there are two types of food allergy, immediate and delayed. While the immediate type may cause a reaction sometimes in minutes, and usually within about two hours, the delayed type, which is caused by a different mechanism, may take up to two days to show.

The susceptibility to allergies tends to run in families. If a particular family is **atopic** (prone to allergies) than it would be recommended that infants are fully breastfed for six months as this reduces the incidence of allergy in later life. A particular problem for babies is the development of allergy to cow's milk as it is not always easy to find satisfactory alternative foods, although goat's milk or soya milk are helpful in some cases. The only real treatment for allergy is for people with hypersensitivity to avoid the offending food. In some cases, where there is only mild sensitivity then heat treatment may render the food harmless.

9.3 Food processing

The words 'food processing' immediately conjure up images of additives in the minds of many consumers, although food scientists would consider the processing effects on reduction of pathogenic organisms to be the issue of paramount importance.

9.4 Function of preservatives and additives

Most consumers in the EC equate preservatives and additivies with E numbers. E numbers indicate that an additive has been through the whole gamut of toxicological tests and accepted at a European level for use in food. The system was introduced to provide a simple method of referring to particular additives without using long chemical names in several languages. While simple to use, the system has created the impression of a large number of substances recently being added to food. In fact, many of the E numbers refer to naturally occurring substances (such as vitamin C, vitamin E, pectin and citric acid) which now appear on ingredients lists with an E number.

Additives are used in foods to improve their physical, sensory (mouthfeel etc.), storage and nutritional characteristics, or as processing aids. Examples of the latter are anti-caking agents to allow free flow of a powder during food manufacture or enzymes (naturally occurring pectinases) added to remove pectin from fruit juice to aid juice extraction. Thickeners such as starches, gums and pectins are used to increase viscosity of products, aerating agents such as baking powder and carbon dioxide (soft drinks) are also added to food. Many of these could be removed from food, but would make food manufacture more expensive and therefore increase the price of the product. Artificial sweeteners are used in foods to lower the energy content and are useful to slimmers and diabetics. As far as artificial colours are concerned, many manufacturers are now offering alternative products, either colour-free or with naturally occurring colours.

Food preservatives often present a 'risk/benefit' problem, that is the risk to health has to be weighed against the benefits to health or of the provision of low-cost food. One very good example is the use of nitrites as a meat preservative.

Nitrites (or **nitrates**) have been used for centuries in the age-old process of curing with brines. We know now that nitrites (nitrates form nitrites in the body) can combine with other substances in the food called amines to produce **nitrosamines**, which are known carcinogens in animals. However, there are several reasons why nitrites or nitrates have not been banned outright in foods:

(*a*) up to 80% of our supply of nitrates and nitrites comes from vegetables in any case and not from cured meat;
(*b*) amounts of nitrosamines formed in foods are very small;
(*c*) there is no epidemiological evidence of cancer in humans, even among people that eat large quantities of cured meats;
(*d*) (and this is where the main 'benefit' is found) nitrite inhibits the growth of the *Clostridium botulinum* spores in canned meat, which would produce the fatal toxin botulin. It is estimated that there would be many more deaths from botulism (see below) if nitrites were banned. Therefore, whereas the effects of nitrosamines are speculative, botulism is a known hazard.

It is true that no government body would be willing to say that any additive was 'risk-free' as it would be impossible to show zero risk of any substance in food; even water is toxic in large amounts. Everything we do in life has risk attached, but there is no doubt that the consumer is much more willing to accept 'voluntary' risk (such as risk of car accident when using that mode of transport) than the 'involuntary' risk of additives in food (seen as a risk imposed from outside). Therefore, many people are not prepared to accept **any** risk from food additives and are demanding 'additive-free' foods, which are now being provided by food manufacturers as alternative products, but often at increased cost.

9.5 Effects of processing on microbiological safety of food

Cold storage and freezing. The growth of mesophiles is retarded by refrigeration, but as storage proceeds there is a change in flora. Thus in milk, souring organisms (mesophiles) are replaced by lipolytic and proteolytic (fat and protein breakdown, respectively) organisms such as *Pseudomonas* or *Achromobacter*. These are psychrophilic organisms, which produce slime and give off unpleasant 'off' odours. The shelf-life of a product during refrigeration depends on the species and numbers of bacteria present before storage. Fish preservation on ice is a good example of this. Fish from the North Sea are high in psychrophiles, which can proliferate at low temperature, so the shelf-life of fresh fish at refrigeration temperature is relatively short. Tropical fish caught in warmer waters, have mainly mesophilic flora, therefore when kept on ice they have a longer shelf-life than temperate fish, as psychrophilic bacteria have to establish themselves.

Any rapid reduction in temperature such as refrigeration, will bring about the death of organisms due to 'cold shock', but some organisms will survive. Death also occurs in frozen storage, but, again, one cannot assume that all organisms will die. In fact frozen storage may only lead to a slow reduction in numbers over a period of months or years and many pathogens will survive freezing, or at least enough of them to cause trouble. This emphasises the need for good quality raw materials for frozen storage.

Heat treatment of foods. Not only do some bacteria (thermophiles) multiply at relatively high temperatures, other bacteria have a special mechanism for survival at even higher temperatures, that is by **spore formation**. The most heat-resistant spores known are produced by certain thermophilic species of *Bacillus*. In canning it is necessary to undertake process calculations, which are based on the D value. The D value (decimal reduction time, for example $10^7 \rightarrow 10^6$ organisms) will depend on the organism which is under consideration.

For canning we normally consider the D value of *Clostridium botulinum* spores, because the toxin produced by these organisms is fatal and as anaerobes they grow well in cans. The cooking required will depend on the loading of bacteria. A heavy load has to be assumed as it is essential that all *C. botulinum* spores are totally eliminated. Thus, the canning industry uses a 12D cook as the basis for calculations for heat processing. Even with a 12D cook, some thermophilic storage bacteria can survive. But these will not proliferate in the can, except when stored under tropical conditions. For this reason cans destined for tropical storage should be given extra cooking.

In general in processing, apart from spoilage considerations, heat treatment of food is aimed at the temperature of destruction of the pathogens present, such as *C. botulinum* for canning and tuberculosis (TB) organisms for the pasteurisation of milk. Although pasteurisation will destroy TB bacteria, some other species survive. Again, the level of organisms present in the starting material is very important as a dirty, pasteurised milk will have a shorter shelf-life than a clean, pasteurised milk. This again emphasises the need for the best quality raw material in food manufacturing.

9.6 Effects of processing on the nutritional value of food

Most people think of the effects of processing on the nutritional value of a food as being detrimental. In fact, as well as the obvious benefits in terms of safer food from a microbiological and toxicological point of view, processing, especially heat treatment, has nutritional benefits which are often overlooked. Protein and starches are made more digestible. This is particularly important for starch, as native starch is very indigestible. Raw potatoes cause scours (diarrhoea) in farm animals and would have the same effect in humans. Heating can make some nutrients more available, as heat breaks down plant intercellular structures (pectin) and allows digestive enzymes greater access to cell contents. Thus beta carotene is more available from cooked carrots than raw carrots.

As far as nutritional losses during processing are concerned, the significance of these will depend on whether the person is marginally provided with a certain nutrient in the total diet. If nutrients are oversupplied, then small losses will be inconsequential. But as we have seen from Chapter 8 there are some people who are at risk of malnutrition and it is these groups which should be particularly borne in mind when considering the effects of processing. For example, particular care should be taken in institutional catering to minimise nutritional losses.

One of the greatest effects on the nutritional value of food is milling of cereals to reduce bran content. The reason for this is that cereals normally provide a large part of the diet and during milling there is considerable loss of dietary fibre (not strictly a nutrient, but important for proper functioning of the gut: see Chapter 6). In addition, there is much loss of calcium, iron and B complex vitamins. Losses of the major components of foods, that is protein, fat and carbohydrate, are of little consequence in the overall diet of an individual. Vitamins are also lost in the preparation of fruit and vegetables, but the extent of this loss depends on the vitamin concerned.

Scientists have attempted to predict nutrient loss from foods during processing, but have found it impossible to do so with any precision because the amounts in food varies (effect of variety and biological variability), the presence of components such as sugar may protect nutrients against processing loss and the pH, which will have a great effect on the outcome of processing, varies. Nevertheless, a knowledge of the factors affecting the stability of vitamins can help in determining which process will have the greatest effect on which vitamin (see Table 9.3). Thus, for example, thiamin (B_1) is particularly affected by heat and therefore would be expected to be lost in canning. Vitamin C is not as heat-labile as is commonly thought. It is, however, very rapidly broken down by ascorbase, an enzyme also found in the plant tissue alongside it but separated from it by cell walls in the intact food. On cutting fruit and vegetables, the enzyme gains access to the vitamin C and breakdown is very rapid. To prevent loss of vitamin C in vegetables, they should be washed before cutting (being water-soluble, vitamin C will leach from the cut surfaces), cutting should be kept to a minimum, and the prepared vegetables should be plunged directly into boiling water which will destroy the enzyme.

Table 9.3 Factors affecting the stability of certain vitamins in food

Vitamin	Water solubility	Subject to oxidation	Heat-labile	Light-sensitive
Vitamin A	no	yes	no	slight
Riboflavin	yes	no	no	yes
Thiamin	yes	no	yes	no
Vitamin C	yes	yes	no	slight

Figure 9.1 gives a rough guide to losses of vitamins which might be expected during cooking, but much will depend on the cooking conditions used.

Vitamin C is the first vitamin in vegetables to be affected by processing and therefore is used to monitor processing change. The vitamin C content of vegetables decreases after harvest quite rapidly and therefore frozen vegetables

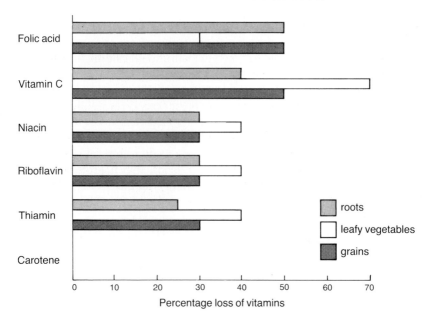

Figure 9.1 Estimated losses of vitamins (%) in plant foods during cooking

can have a greater content than those purchased 'fresh', particularly if the latter have spent some time in the wholesale and retail chains.

With the recent introduction of **microwave ovens** into many homes, consumers are naturally concerned about the losses of vitamins compared with conventional cooking. The only advantage over conventional cooking methods is that little water is needed for microwave cooking and therefore there is markedly less leaching. Thus, frozen peas cooked in the microwave with no water contain more vitamin C than those cooked conventionally. If the same amount of water is used as in conventional cooking then losses of vitamin C are very similar.

9.7 Conclusion

Food processing provides the means of reducing microbiological and toxicological hazard. On an industrial scale, these effects are constantly monitored, with the result that we have come to rely on the supply of safe food and even take it for granted. Perhaps because of this, consumer concern has focused, until recently, on additives and the nutritional value of processed food. This has led to consumer pressure on manufacturers to produce a wide range of alternative products including additive-free foods. Recent food poisoning outbreaks have focused consumer interest on the microbiological hazard of food demonstrating that the issue is not historical, that the microorganisms are still with us and that we should not become complacent on this issue.

Although much food eaten today is already processed on purchase, and some loss of nutrients is inevitable, losses are small. Well-nourished, healthy people are unaffected by small losses in nutrients during processing, as the diet will normally provide these in excess of requirements. However, nutrient losses may become significant as the amounts in the diet approach requirement levels.

Vulnerable groups, such as children, pregnant and lactating women, old people, sick and convalescent people, and those on restricted diets (such as vegetarians) may be at risk of not obtaining all nutrients required. Therefore, particular attention should be paid to their diets and the supply of certain nutrients for them. Specific dietary problems in a developed country such as the UK (such as isolated pockets of rickets) would be better approached through individual advice and education rather than a 'blanket' approach, such as food fortification. In the UK well-processed food may not be leading to problems of undernutrition, but rather to overnutrition, with food that is too tasty and available in too much variety!

Further reading

Board, R.G. (1983) *Modern Introduction to food microbiology*, Blackwells

Cornwell, A. and Cornwell, V. (1987) *Drugs, alcohol and mental health*, Social Biology Series, Cambridge University Press

Department of Health (1991) *Dietary Reference Values for food energy and nutrients for the United Kingdom*, Report on Health and Social Subjects No. 41, HMSO

DHSS (Department of Health and Social Security, Great Britain) (1979) *Recommended daily amounts of food energy and nutrients for groups of people in the United Kingdom*, Report on Health and Social Subjects No. 15, HMSO

DHSS (1984) *Diet and cardiovascular disease* (The *COMA report*), Report on Health and Social Subjects No. 28, HMSO

DHSS (1988) *Present day practice in infant feeding: third report*, Report on Health and Social Subjects No. 32, HMSO

FAO (1973) *Energy and protein requirements*, FAO Nutrition Meetings Report Series No. 52, Food and Agriculture Organisation of the United Nations, Rome

Farrer, K.T.H. (1987) *A guide to food additives and contaminants*, The Parthenon Publishing Group

Fox, B.A. and Cameron, A.G. (1989) *Food science, nutrition and health.* 5th ed., Edward Arnold

Gibney, M.J. (1986) *Nutrition, diet and health*, Cambridge University Press

Gunstone, F.D. and Norris, F.A. (1983) *Lipids in foods: chemistry, biochemistry and technology*, Pergamon

Guthrie, H.A. (1983) *Introductory nutrition*, 5th ed., C.V. Mosby Co.

Heimann, W. (1980) *Fundamentals of food chemistry*, Ellis Horwood

Jay, J.M. (1986) *Modern food microbiology*, 3rd ed., Van Nostrand Reinhold

Jukes, D.J. (1987) *Food legislation in the UK: a concise guide*, 2nd ed., Butterworths

Ministry of Agriculture, Fisheries and Food (1985) *Manual of nutrition*, 9th ed., HMSO

NACNE (National Advisory Committee on Nutrition Education) (1983) *Proposals for nutritional guidelines for health education in Britain* (The *NACNE report*), The Health Education Council, New Oxford St., London WC1

Passmore, R. and Eastwood, M.A. (1986) *Davidson and Passmore: Human nutrition & dietetics*, 8th ed., Churchill Livingstone

Paul, A.A. and Southgate, D.A.T. (1978) *McCance and Widdowson's The composition of foods*, 4th ed., MRC Special Report No. 297, HMSO

Pike, R.L. and Brown, M.L. (1984) *Nutrition: an integrated approach*, 3rd ed., John Wiley and Sons

Pyke, M. (1975) *Success in nutrition*, John Murray

Pyke, M. (1981) *Food science and technology*, 4th ed., John Murray

Appendix 1

Dietary Reference Values (DRVs) for food energy and selected nutrients for population groups in the United Kingdom

EARs → Reference Nutrient Intakes (RNIs) →

Males:	Energy MJ/d	(kcal/d)	Protein[a] g/d	Thiamin mg/d	Riboflavin mg/d	Niacin (nicotinic acid equivalent) mg/d	Folate μg/d	Vitamin C mg/d	Vitamin A μg/d	Vitamin D μg/d	Calcium mg/d	Iron mg/d	Zinc mg/d
0–3 months	2.28	(545)	12.5	0.2	0.4	3	50	25	350	8.5	525	1.7	4.0
4–6 months	2.89	(690)	12.7	0.2	0.4	3	50	25	350	8.5	525	4.3	4.0
7–9 months	3.44	(825)	13.7	0.2	0.4	4	50	25	350	7	525	7.8	5.0
10–12 months	3.85	(920)	14.9	0.3	0.4	5	50	25	350	7	525	7.8	5.0
1–3 years	5.15	(1,230)	14.5	0.5	0.6	8	70	30	400	7	350	6.9	5.0
4–6 years	7.16	(1,715)	19.7	0.7	0.8	11	100	30	500	—	450	6.1	6.5
7–10 years	8.24	(1,970)	28.3	0.7	1.0	12	150	30	500	—	550	8.7	7.0
11–14 years	9.27	(2,220)	42.1	0.9	1.2	15	200	35	600	—	1,000	11.3	9.0
15–18 years	11.51	(2,755)	55.2	1.1	1.3	18	200	40	700	—	1,000	11.3	9.5
19–50 years	10.60	(2,550)	55.5	1.0	1.3	17	200	40	700	—	700	8.7	9.5
51–59 years	10.60	(2,550)	53.3	0.9	1.3	16	200	40	700	—	700	8.7	9.5
60–64 years	9.93	(2,380)	53.3	0.9	1.3	16	200	40	700	—	700	8.7	9.5
65–74 years	9.71	(2,330)	53.3	0.9	1.3	16	200	40	700	10	700	8.7	9.5
75+ years	8.77	(2,100)	53.3	0.9	1.3	16	200	40	700	10	700	8.7	9.5

Females:

0–3 months	2.16	(515)	12.5^b	0.2	0.4	3	50	25	350	8.5	525	1.7	4.0
4–6 months	2.69	(645)	12.7	0.2	0.4	3	50	25	350	8.5	525	4.3	4.0
7–9 months	3.20	(765)	13.7	0.2	0.4	4	50	25	350	7	525	7.8	5.0
10–12 months	3.61	(865)	14.9	0.3	0.4	5	50	25	350	7	525	7.8	5.0
1–3 years	4.86	(1,165)	14.5	0.5	0.6	8	70	30	400	7	350	6.9	5.0
4–6 years	6.46	(1,545)	19.7	0.7	0.8	11	100	30	500	—	450	6.1	6.5
7–10 years	7.28	(1,740)	28.3	0.7	1.0	12	150	30	500	—	550	8.7	7.0
11–14 years	7.92	(1,845)	41.2	0.7	1.1	12	200	35	600	—	800	14.8^b	9.0
15–18 years	8.83	(2,110)	45.0	0.8	1.1	14	200	40	600	—	800	14.8^b	7.0
19–50 years	8.10	(1,940)	45.0	0.8	1.1	13	200	40	600	—	700	14.8^b	7.0
51–59 years	8.00	(1,900)	46.5	0.8	1.1	12	200	40	600	—	700	8.7	7.0
60–64 years	7.99	(1,900)	46.5	0.8	1.1	12	200	40	600	—	700	8.7	7.0
65–74 years	7.96	(1,900)	46.5	0.8	1.1	12	200	40	600	10	700	8.7	7.0
75+ years	7.61	(1,810)	46.5	0.8	1.1	12	200	40	600	10	700	8.7	7.0
Pregnancy^c	+0.80	(200)	+6	+0.1**	+0.3	*	+100	+10	+100	10	*	*	*
Lactation^c													
0–4 months	≤ {+2.40	(570)	+11	+0.2	+0.5	+2	+60	+30	+350	10	+550	*	+6.0
4+ months	{+2.40	(570)	+8	+0.2	+0.5	+2	+60	+30	+350	10	+550	*	+2.5

EAR, Estimated Average Requirement (see Figure 1.2): *a*, based on a good quality protein; *b*, insufficent if menstrual loss is high, when supplements are suggested: *c*, to be added to adult requirement; *, no increment; **, for last trimester only.

From: Department of Health (1991)

Index

absorption **28**
additives **86**
adrenalin *see endocrine glands*
aflatoxin **84**
alcohol 73, 77
alimentary canal **22**
allergies 83, **85**
amino acids, 15
   deamination **35**, 36
   transamination **35**
anaemia, iron deficiency 51, 63, **77**
anorexia nervosa **75**
antioxidants 21
atherosclerosis 53, 71, 72
Atwater factors 48

beans *see legumes*
bioavailability of minerals 30, 59
body composition **4**

calcium
   absorption 29
   balance **51**, 63
   deficiency 52, 63, 66, 67, 73
   recommended daily amount (RDA) 64, 94
   sources 67, 89
cancer 57, 69, 84
carbohydrates **11**
   absorption **28**
   cellulose 13, 14
   digestion **25**
   disaccharides 12
   hemicellulose 13
   metabolism **31**
   monosaccharides 11
   sugars 11
carotenes 53, 88

cereals 67, 68, 78, 89
cholesterol 19, 55, 60, 72
composition of foods' tables 8
cyanogens **84**

diabetes 43, 57, **68**, 69, 71
diet
   adequacy of **63**
   fat in 10
   starch in 10
   sugar in 10
dietary fibre 13, **55**, 89
   analysis **57**
   bioavailability of minerals **59**
   definition **57**
   nutritional value **58**
   physiological effects **59**, 67, 73
   satiety 59
   sources **58**
dietary guidelines **9**, 58, 73
dietary surveys **8**, 63, 75
digestion **22**
   carbohydrates **25**
   gastrointestinal hormones 25, 26
   gut secretions **24**
   lipids **26**
   proteins **27**
diseases of affluence 9
diverticular disease 56, 57

elderly people **66**
endocrine glands **38**
   adrenalin 39, 42, 43
   glucagon 41, 42, 43
   growth hormone 38

hypothalamus **39**
insulin 41, 42, 43
pancreas, endocrine function **41**
pituitary gland **39**
thyroid **38**
   hypothyroidism 39
   goitre 38
   thyrotoxicosis 38
energy
   Atwater factors 48
   expenditure **46**
   intake **46**
   physical exercise 62, 70
   requirements **61**
   sources 45
   units **45**
epidemiology 56, 72
essential fatty acids (EFA) *see fats*
exercise, physical 62, 66, 70, 71

fats
   absorption **28**
   dietary 10
   digestion **26**
   essential fatty acids (EFA) 19, 20, 54, 55
   ketone bodies 36
   metabolism **36**
   P/S ratio 55
   polyunsaturated fatty acids (PUFA) 20, 54, 55, 74
   rancidity 20
   saturated fatty acids 20
   triacylglycerols 19
   triglycerides 19
favism **85**
fibre *see dietary fibre*

folic acid 63, 64
food-borne diseases 79,
     81
     Clostridium botulinum
          83, 87, 88
     Listeria 79, 83
     Salmonella 79, 83
food industry 79
free radicals 21

glucagon *see endocrine
     glands*
glucose, blood
     glucostat function of
          liver 41
     regulation **41**
goitre **76**
growth hormone *see
     endocrine glands*

haemagglutinins **85**
heart disease 57, 69, **71**
homeostasis 16, 38
hormones *see endocrine
     glands*
human milk 64
hypothalamus *see
     endocrine glands*

infant nutrition 63, **64**
     formulas 65, 79
insulin *see endocrine
     glands*
iron
     absorption 29, 67, 78
     balance **50**, 63
     deficiency 51, 63, **77**
     recommended daily
          amount (RDA) 64,
          94
     sources 89

lactation 63, **64**
lectins *see
     haemagglutinins*
legumes 48, 67, 68, 78
liver **40**
     glucostat function 41
lymphatic system 24, 28

malnutrition 63, **69**, *see
     also protein-energy
     malnutrition*

metabolism **31**
     carbohydrate **31**, 42
     citric acid cycle **33**
     protein **34**
     respiratory chain **33**
microorganisms *see food-
     borne diseases*
minerals *see also iron,
     calcium*
     absorption **29**
     balance **50**
     bioavailability 30, 59
     load in infant
          formulas 65

nicotinic acid (niacin)
     29, 63
nutrient balance **6**, **45**
nutrients **5**
     absorption **28**
     deficiency studies **5**
     physiological
          requirements **6**
     recommended daily
          amounts (RDA) 6,
          94
nutrition, definition **3**

obesity **69**

pancreas *see endocrine
     glands*
pathogens *see food-borne
     diseases*
pellagra 29, 63
phytate 30, 59, 78
pituitary gland *see
     endocrine glands*
pregnancy 63, **64**
preservatives **86**
processing food 81, **86**,
     87
     freezing **87**
     heat treatment **88**
prostaglandins 55
protein energy
          malnutrition 62,
          63, 66, **74**
protein quality **18**
proteins
     absorption **28**
     digestion **27**
     metabolism **34**

requirements **49**
structure **15**

recommended daily
     amount (RDA) of
          nutrients **6**, 63, 64,
          75, **94**
riboflavin 39, 63
rickets 52

salt (sodium) 73
slimming *see weight
     reduction*
starch **13**
     dietary 10
     effects of processing
          88
starvation 36, 42
sucrose *see sugars*
sugars **11**, 73

thiamin 39, 63, 89
thrombus formation 71,
     74
thyroid *see endocrine
     glands*
toxins in foods 41, **83**
trypsin 27

urea cycle 36

vegetarians 63, **67**, 78
vitamins **52**
     absorption **29**
     intake 63
     stability 89
     vitamin A **53**
          beta carotene 53
          deficiency 63, **76**
     vitamin B$_{12}$ 39, 67
     vitamin C **52**, 63, 64,
          66, 67, 77, 89, 90
     vitamin D 30, 63, 66,
          67

water **50**
weaning 66
weight reduction 45, 59,
     70

xerophthalmia 63, **76**

Zinc 63